COMPANION GUIDE TO INFECTIOUS DISEASES OF MICE AND RATS

Committee on Infectious Diseases of Mice and Rats
Institute of Laboratory Animal Resources
Commission on Life Sciences
National Research Council

NATIONAL ACADEMY PRESS
Washington, D. C. 1991

NATIONAL ACADEMY PRESS • 2101 CONSTITUTION AVE., N.W. • WASHINGTON, D.C. 20418

NOTICE: The project that is the subject of this report was approved by the Governing Board of the National Research Council, whose members are drawn from the councils of the National Academy of Sciences, National Academy of Engineering, and Institute of Medicine. The members of the committee responsible for the report were chosen for their special competences and with regard for appropriate balance.

This report has been reviewed by a group other than the authors according to procedures approved by a Report Review Committee consisting of members of the National Academy of Sciences, National Academy of Engineering, and Institute of Medicine.

The National Academy of Sciences is a private, nonprofit, self-perpetuating society of distinguished scholars engaged in scientific and engineering research, dedicated to the furtherance of science and technology and to their use for the general welfare. Upon the authority of the charter granted to it by the Congress in 1863, the Academy has a mandate that requires it to advise the federal government on scientific and technical matters. Dr. Frank Press is president of the National Academy of Sciences.

The National Academy of Engineering was established in 1964, under the charter of the National Academy of Sciences, as a parallel organization of outstanding engineers. It is autonomous in its administration and in the selection of its members, sharing with the National Academy of Sciences the responsibility for advising the federal government. The National Academy of Engineering also sponsors engineering programs aimed at meeting national needs, encourages education and research, and recognizes the superior achievements of engineers. Dr. Robert M. White is president of the National Academy of Engineering.

The Institute of Medicine was established in 1970 by the National Academy of Sciences to secure the services of eminent members of appropriate professions in the examination of policy matters pertaining to the health of the public. The Institute acts under the responsibility given to the National Academy of Sciences by its congressional charter to be an adviser to the federal government and upon its own initiative to identify issues of medical care, research, and education. Dr. Samuel O. Thier is president of the Institute of Medicine.

The National Research Council was established by the National Academy of Sciences in 1916 to associate the broad community of science and technology with the Academy's purposes of furthering knowledge and advising the federal government. Functioning in accordance with general policies determined by the Academy, the Council has become the principal operating agency of both the National Academy of Sciences and National Academy of Engineering in the conduct of their services to the government, the public, and the scientific and engineering communities. The Council is administered jointly by both Academies and the Institute of Medicine. Dr. Frank Press and Dr. Robert M. White are chairman and vice-chairman, respectively, of the National Research Council.

This project was supported by the National Cancer Institute, National Institutes of Health, under contracts NO1-CM-57644 and NO1-CM-07316, administered by the Division of Cancer Treatment.

Library of Congress Cataloging-in-Publication Data

Companion guide to infectious diseases of mice and rats / Committee on Infectious Diseases of Mice and Rats, Institute of Laboratory Animal Resources, Commission on Life Sciences, National Research Council.
    p. cm.
  Includes bibliographical references.
  ISBN 0-309-04283-6
  1. Mice–Infections. 2. Rats–Infections. 3. Laboratory animals–Infections. 4. Mice as laboratory animals. 5. Rats as laboratory animals. I. Institute of Laboratory Animal Resources (U. S.). Committee on Infectious Diseases of Mice and Rats. II. Infectious diseases of mice and rats.
SF996.5.C65 1991
636'.93233–dc20   91-10495
                                                                                        CIP

Copyright © 1991 by the National Academy of Sciences

No part of this book may be reproduced by any mechanical, photographic, or electronic process, or in the form of a phonographic recording, nor may it be stored in a retrieval system, transmitted, or otherwise copied for public or private use, without written permission from the publisher, except for the purposes of the official use of the U.S. government.

Printed in the United States of America

COMMITTEE ON INFECTIOUS DISEASES OF MICE AND RATS

J. Russell Lindsey (Chairman), Department of Comparative Medicine, Schools of Medicine and Dentistry, University of Alabama at Birmingham, and Birmingham Veterans Administration Medical Center

Gary A. Boorman, Chemical Pathology Branch, Toxicological Research and Testing Program, National Institute of Environmental Health Sciences, Research Triangle Park, North Carolina

Michael J. Collins, Jr., Animal Health Diagnostic Laboratory, NCI-Frederick Cancer Research Facility, Frederick, Maryland

Chao-Kuang Hsu, Smith Kline Animal Health Products, West Chester, Pennsylvania

Gerald L. Van Hoosier, Jr., Department of Comparative Medicine, University of Washington, Seattle

Joseph E. Wagner, Veterinary Medical Diagnostic Laboratory, College of Veterinary Medicine, University of Missouri-Columbia

Staff

Dorothy D. Greenhouse, Senior Program Officer

The Institute of Laboratory Animal Resources (ILAR) was founded in 1952 under the auspices of the National Research Council. Its mission is to provide expert counsel to the federal government, the biomedical research community, and the public on the scientific, technological, and ethical use of laboratory animals within the context of the interests and mission of the National Academy of Sciences. ILAR promotes the high-quality humane care of laboratory animals; the appropriate use of laboratory animals; and the exploration of alternatives in research, testing, and teaching.

INSTITUTE OF LABORATORY ANIMAL RESOURCES COUNCIL

Steven P. Pakes (Chairman), The University of Texas Southwestern Medical Center, Dallas
June R. Aprille, Tufts University, Medford, Massachusetts
Melvin W. Balk, Charles River Laboratories, Inc., Wilmington, Massachusetts
Douglas M. Bowden, University of Washington, Seattle
Lester M. Crawford, U.S. Department of Agriculture, Washington, D.C.
Thomas J. Gill III, University of Pittsburgh School of Medicine, Pittsburgh, Pennsylvania
Alan M. Goldberg, The Johns Hopkins University, Baltimore, Maryland
Jon W. Gordon, Mt. Sinai School of Medicine, New York, New York
Margaret Z. Jones, Michigan State University, East Lansing
Michael D. Kastello, Merck Sharp & Dohme Research Laboratories, Rahway, New Jersey
Robert H. Purcell, National Institute of Allergy and Infectious Diseases, Bethesda, Maryland
J. Wesley Robb, School of Medicine, University of Southern California, Los Angeles
John L. VandeBerg, Southwest Foundation for Biomedical Research, San Antonio, Texas

Staff
Thomas L. Wolfle, Director

COMMISSION ON LIFE SCIENCES

Bruce M. Alberts (Chairman), University of California, San Francisco
Bruce N. Ames, University of California, Berkeley
Francisco J. Ayala, University of California, Irvine
J. Michael Bishop, University of California Medical Center, San Francisco
Michael T. Clegg, University of California, Riverside
Glenn A. Crosby, Washington State University, Pullman
Freeman J. Dyson, The Institute for Advanced Study, Princeton, New Jersey
Leroy E. Hood, California Institute of Technology, Pasadena
Donald F. Hornig, Harvard University School of Public Health, Boston, Massachusetts
Marian E. Koshland, University of California, Berkeley
Richard E. Lenski, University of California, Irvine
Steven P. Pakes, The University of Texas Southwestern Medical Center, Dallas

Emil A. Pfitzer, Hoffmann-LaRoche, Inc., Nutley, New Jersey
Thomas D. Pollard, The Johns Hopkins University, Baltimore, Maryland
Joseph E. Rall, National Institutes of Health, Bethesda, Maryland
Richard D. Remington, University of Iowa, Iowa City
Paul G. Risser, University of New Mexico, Albuquerque
Harold M. Schmeck, Jr., Armonk, New York
Richard B. Setlow, Brookhaven National Laboratory, Upton, New York
Carla J. Shatz, Stanford University School of Medicine, Stanford, California
Torsten N. Wiesel, Rockefeller University, New York, New York

Staff
John E. Burris, Executive Director

Preface

This handbook is a companion to the volume *Infectious Diseases of Mice and Rats*. It summarizes the information in the longer text and is intended to serve as a guide for biomedical scientists and for veterinarians and others associated with an animal resources program to assist them in identifying infectious agents of mice and rats and determining the effect of these agents on their research. Like *Infectious Diseases of Mice and Rats*, the handbook is comprised of three parts: Part I, Principles of Rodent Disease Prevention, summarizes basic concepts and practices for detecting and excluding infectious diseases from animal facilities; Part II, Disease Agents, provides pertinent information on the epizootiology, pathogenesis, diagnosis, and control of infectious agents and the effects of these agents on research; and Part III, Diagnostic Indexes, contains tabular information intended as an aid to diagnostic problem solving.

The committee extends its thanks to the staff of the Institute of Laboratory Animal Resources, which worked with the committee to summarize the information in *Infectious Diseases of Mice and Rats* to produce this guide.

J. Russell Lindsey, Chairman
Committee on Infectious Diseases of Mice and Rats

Contents

PART I: PRINCIPLES OF RODENT DISEASE PREVENTION 1
Scientific Objectives 1
Infection Versus Disease 1
Terminology of Microbial and Pathogen Status 2
Commitment to Maintaining Pathogen-Free Status of Rodents 2
Health Surveillance Programs 3
Rodent Diagnostic Laboratories 6
References 7

PART II: DISEASE AGENTS 9
Bacteria, Fungi, and Viruses 9
  Adenoviruses 9
  Bacillus piliformis 10
  Cilia-Associated Respiratory Bacillus 12
  Citrobacter freundii Biotype 4280 13
  Corynebacterium kutscheri 14
  Cytomegalovirus, Mouse 15
  Ectromelia Virus 16
  Encephalitozoon cuniculi 18
  Hantaviruses 20
  Hepatitis Virus, Mouse 21
  H-1 Virus 24
  Kilham Rat Virus 25
  Lactic Dehydrogenase-Elevating Virus 26
  Leukemia Viruses, Murine 28
  Lymphocytic Choriomeningitis Virus 30

Mammary Tumor Virus, Mouse 32
Minute Virus of Mice 33
Mycoplasma arthritidis 35
Mycoplasma pulmonis 36
Pasteurella pneumotropica 38
Pneumocystis carinii 39
Pneumonia Virus of Mice 40
Polyomavirus 41
Pseudomonas aeruginosa 42
Reovirus-3 43
Rotavirus, Mouse 44
Salmonella enteritidis 45
Sendai Virus 47
Sialodacryoadenitis Virus 50
Staphylococcus aureus 51
Streptobacillus moniliformis 53
Streptococcus pneumoniae 54
Theiler's Murine Encephalomyelitis Virus 55
Thymic Virus, Mouse 56

Dermatophytes 57
Trichophyton spp. and *Microsporum* spp. 57

Common Ectoparasites 58
Myobia musculi 58
Myocoptes musculinus and *Radfordia affinis* 60
Other Ectoparasites 60

Endoparasites 61
Aspicularis tetraptera (Mouse Pinworm) 61
Entamoeba muris 61
Giardia muris 62
Hymenolepis nana 64
Spironucleus muris 65
Syphacia obvelata (Mouse Pinworm) and *Syphacia muris* (Rat Pinworm) 66
Trichomonas muris 67
Other Endoparasites 68

PART III: INDEXES TO DIAGNOSIS AND RESEARCH COMPLICATIONS OF INFECTIOUS AGENTS 70
Introduction 70
Clinical Signs 71
Pathology 76
Research Complications 84

INDEX 87

PART I

Principles of Rodent Disease Prevention

SCIENTIFIC OBJECTIVES

Animal experiments are essential to progress in the biomedical sciences (NRC, 1985). Like investigations in any field of science, the merit of animal experiments ultimately depends on rigid adherence to the principles of scientific method. Proper practice of these principles yields data that are both reliable and reproducible, key objectives of all good experiments (Bernard, 1865).

INFECTION VERSUS DISEASE

A common misconception is that infection is synonymous with disease. Bacterial opportunists and commensals, which are constitutents of the normal flora on mucosal and body surfaces, are ubiquitous infections that usually cause disease only when their hosts are immunosuppressed (Dubos et al., 1965; Savage, 1971). The viral and parasite pathogens of rodents vary considerably in pathogenicity. Some cause severe disease, while others rarely do. It is also important to distinguish between subclinical (inapparent, covert, or silent) and clinically apparent infections. Most natural infections with pathogenic organisms in mice and rats are subclinical, and infection-induced aberrations in research results often occur in the absence of clinical disease. Thus, it is important to prevent infection, not merely to prevent clinical disease.

TERMINOLOGY OF MICROBIAL AND PATHOGEN STATUS

Terms used in defining rodent microbial status vary greatly in precision of meaning. Four terms (germfree, gnotobiotic, defined flora, and conventional), representing the extremes of microbial status, have clear definitions that are generally accepted and understood by scientists, as well as by technical personnel (NRC, 1991). However, there is major confusion about the definition and use of terms representing the middle ground of pathogen status. Pathogen free, specific pathogen free, virus antibody free, and clean conventional are relative terms that require explicit definition every time they are used. The definition should include the background of the rodent subpopulation in question (e.g., cesarean derived, isolator maintained, barrier maintained), details of current housing (e.g., isolator, barrier), and data from laboratory tests for pathogens (the specific tests done, the number of tests, the frequency of testing, and the results) (Lindsey et al., 1986).

COMMITMENT TO MAINTAINING PATHOGEN-FREE STATUS OF RODENTS

Past experience demonstrates that maintaining rodents in the pathogen-free state requires adherence to breeding, transportation, and maintenance programs specifically designed for the exclusion of pathogens. This means a strong commitment by investigators, research staff, and animal care staff. Some essential elements of that commitment are as follows:*

   a. The investigator and the support personnel must understand the terminology and principles involved.

   b. Appropriate facilities and equipment must be available.

   c. Housing practices must ensure physical separation and avoidance of cross-contamination between different animal subpopulations throughout their lives.

   d. Reliable health monitoring should be maintained to identify breeding populations free of pathogens and to redefine the microbiologic status of the animals at regular intervals from the time they are received in the user facility until completion of each study.

   e. Written standard operating practices must be developed and followed without interruption; clear objectives must be defined in advance, along with detailed procedures for reaching those objectives.

---

*From a consensus developed during a seminar entitled "Barrier Maintenance of Rodents in Multipurpose Facilities," held at the Thirty-Sixth Annual Session of the American Association for Laboratory Animal Science on November 3-8, 1985, in Baltimore, Md. Participants were J. R. Lindsey (leader); G. L. Van Hoosier, Jr.; D. B. Casebolt; J. G. Fox; R. O. Jacoby; and T. E. Hamm, Jr.

HEALTH SURVEILLANCE PROGRAMS

Health surveillance (or monitoring) is the term usually applied to the testing of laboratory animals to determine their pathogen status and general state of health. Health surveillance programs are systematic laboratory investigations that employ batteries of diagnostic tests for the purpose of defining the pathogen and health status of an animal population. These programs are crucially important in rodent disease prevention because they provide data, which are the only reliable basis for determining rodent pathogen status or providing health quality assurance.

Although the need for health surveillance programs is generally accepted, there is a great diversity of opinion about the design of individual programs (Hsu et al., 1980; Iwai et al., 1980; Thigpen and Tortorich, 1980; Jacoby and Barthold, 1981; Hamm, 1983; Loew and Fox, 1983; Small, 1984). No two programs are identical. Some are limited in scope; others are very comprehensive. Numerous factors should be considered in designing individual programs, and special emphasis should be placed on objectivity in testing rather than on the adoption of customary practices. Some of those factors are listed in the following sections.

Scientific Objectives

Health surveillance efforts should, to the fullest extent possible, be matched qualitatively and quantitatively with the specific scientific objectives of individual research programs to ensure that the quality of the animal will meet these objectives. For practical reasons, it is impossible to test for all known infectious agents of rodents, or even all infectious agents that theoretically could interfere with a particular study. In designing health surveillance programs, decisions must be made about which agents should be covered in the test battery. Inclusion of a pathogen in the test battery should be based on the likelihood that it will interfere with the research being conducted. Such information is given in Part II of this volume.

Test Procedures

The procedures used in health surveillance generally include serologic tests, bacterial cultures, parasitologic examinations, and histopathology. Each category can include a few or many procedures to detect different infectious agents or disease processes. Some health surveillance programs are limited to only one of these types of procedures, e.g., serologic testing.

Serologic tests are the main procedures used for detecting virus infections in rodents, but they also have been found useful for some bacterial and protozoan infections. The enzyme-linked immunosorbent assay (ELISA) and the indirect immunofluorescent antibody (IFA) test have largely replaced the complement fixation (CF) test and the hemagglutination inhibition (HAI) test. They are much more sensitive than either the CF or HAI test and give fewer false positives than the

HAI test. Serologic testing should rely on a primary test for each agent and one or more additional tests to confirm the positive results of any primary test (Kraft and Meyer, 1986; Smith, 1986b; Van Der Logt, 1986).

One of the most useful applications of serologic testing in rodent health surveillance is the mouse antibody production (MAP) test (Rowe et al., 1959, 1962). Although originally developed as a method for broadly screening mouse tissues for viruses, it can be used to test transplantable tumors, hybridomas, cell lines, and other biologic materials for contamination by infectious agents. An equivalent test, the rat antibody production (RAP) test, is useful for screening biologic materials taken from rats. Both these tests are generally considered more sensitive than virus isolation (de Souza and Smith, 1989).

The isolation of bacteria using cultural methods and the demonstration and identification of parasites using a microscope are the standard procedures for detection of these agents. However, these methods also have limitations, depending on the agent. In general, causative agents are more difficult to isolate or demonstrate in subclinical infections than in clinically apparent infections. Some bacterial infections of mice and rats, e.g., *Corynebacterium kutscheri* or *Mycoplasma arthritidis*, commonly occur as subclinical infections in which cultural isolation is extremely difficult. With each of the bacteria and parasites it is imperative that specimens be collected from the most appropriate site(s) and processed expeditiously using methods known to maximize the chances of successful recovery or demonstration of the agent. Failure to collect specimens from the site that is most appropriate for that microbe can result in false-negative tests.

Gross and microscopic evaluations of tissues for lesions are also invaluable in health surveillance. In more comprehensive health surveillance programs, histopathologic examination of all major organs by a qualified pathologist is standard practice. Lesions caused by viral pathogens can occur before seroconversion. Some histopathologic changes are diagnostic; others provide only clues to disease processes.

Diagnostic methodology is in transition. Refinements continue to be made in existing methods, and newer methods employing molecular biologic techniques, e.g., nucleic acid hybridization and specific gene product detection, are being developed at a rapid pace (Sklar, 1985; Edberg, 1986; Smith, 1986a,c; Delellis and Wolfe, 1987; Howanitz, 1988).

Sampling Strategies

The purpose of health surveillance is to detect at least one animal with each of the infections or diseases present in the population. The purpose is not to determine prevalence of infection or disease.

The number of animals (sample size) to be tested is of critical importance and can be determined mathematically by making important assumptions about the rates of infection and the randomness in sampling (ILAR, 1976; Hsu et al., 1980; Small,

TABLE 1 Confidence Limits for Detecting Infection Using Different Sample Sizes and Assumed Rates of Infection[a]

Sample Size (N)[b]	Assumed Infection Rate (%)										
	1	2	3	4	5	10	15	20	25	30	40	50
5	0.05	0.10	0.14	0.18	0.23	0.41	0.56	0.67	0.76	0.83	0.92	0.97
10	0.10	0.18	0.26	0.34	0.40	0.65	0.80	0.89	0.94	0.97	0.99	
15	0.14	0.26	0.37	0.46	0.54	0.79	0.91	0.95	0.99			
20	0.18	0.33	0.46	0.56	0.64	0.88	0.95	0.99				
25	0.22	0.40	0.53	0.64	0.72	0.93	0.98					
30	0.25	0.45	0.60	0.71	0.79	0.96	0.99					
35	0.30	0.51	0.66	0.76	0.83	0.97						
40	0.33	0.55	0.70	0.80	0.87	0.99						
45	0.36	0.69	0.75	0.84	0.90	0.99						
50	0.39	0.64	0.78	0.87	0.92	0.99						
60	0.45	0.70	0.84	0.91	0.95							
70	0.51	0.76	0.88	0.94	0.97							
80	0.55	0.80	0.91	0.96	0.98							
90	0.60	0.84	0.94	0.97	0.99							
100	0.63	0.87	0.95	0.98	0.99							
120	0.70	0.91	0.97	0.99								
140	0.76	0.94	0.99									
160	0.80	0.96	0.99									
180	0.84	0.97										
200	0.87	0.98										

[a] From ILAR (1976a), Hsu et al. (1980), and Small (1984).

[b] $N = \dfrac{\log(1 - \text{probability of detecting infection})}{\log(1 - \text{assumed infection rate})}$

1984; DiGiacomo and Koepsell, 1986). As shown in Table 1, if one assumes that 40% of the animals in a population are infected with an agent, there is a 99% probability that 1 infected animal will be detected in a randomly selected sample of 10 animals. At a 50% infection rate, a sample size of only 5 is required for a 97% probability of detecting infection in at least 1 animal.

Although the sample size required to detect a single agent can be determined with reasonable precision, it is virtually impossible to maintain the same degree of precision for all agents to be included in a large test battery. Different agents typically have very different infection rates within rodent colonies. For example, typical rates for established infections in mouse colonies are greater than 90% for Sendai virus, approximately 25% for pneumonia virus of mice, and less than 5% for *Salmonella enteritidis*. In determining the number of animals to be used in a health surveillance test battery for these three agents, the lowest assumed infection rate should be used (i.e., 5%), and a 95% confidence limit would require a sample size of at least 60 animals. This is entirely appropriate in instances where subclinical *S. enteritidis* infection is suspected. However, for routine health surveillance, sample sizes are usually based on assumed infection rates of 40-50% in order to keep sample sizes reasonable.

Proper sampling also requires that animals be taken from different cages, shelves, and racks so that the sample is representative of the entire population. Animals of both sexes and of two age groups should be sampled. For serologic testing, sampling of young adults (approximately 90 days old) and retired breeders is recommended. Young adults are best for detecting recent viral infections (without interference from passive antibody), and retired breeders give an indication of the colony's infection history (Jacoby and Barthold, 1981).

Test Frequency

One of the most difficult decisions to be made in designing health surveillance programs is how frequently a given rodent population should be tested. There are no established guidelines, but the problem seems to revolve around four central issues: the specific purpose of the population in question, the potential or real importance of a pathogen or other contamination to use of the population, the level of risk of pathogen contamination from other nearby rodent populations, and economic considerations. After evaluating these issues, one should have a basis for deciding whether testing should be monthly, quarterly, biannually, or annually. However, the frequency of testing may be different for different agents. For example, if the greatest risks are deemed to be from mouse hepatitis virus and Sendai virus, tests for these agents could be performed monthly, and a larger battery could be done biannually.

Sentinel Animals

Sentinel rodents are sometimes introduced into a rodent population, housed in open cages placed systematically throughout the colony, and used periodically for testing. Pathogen transmission from the principal population to the sentinels can be increased by transferring the sentinels into dirty cages from the principal population at each cage change. Sentinel animals preferably should be of the same population as the principal population and should be subjected to any experimental treatments given to the principal population. The introduction of a second population as sentinels, even if it is tested and found to be free of pathogens, poses an unnecessary risk for contaminating the principal population.

RODENT DIAGNOSTIC LABORATORIES

Rodent diagnostic laboratories are indispensable to the production and maintenance of mice and rats for high-quality research. Such laboratories specialize in health surveillance testing, investigations of clinical diseases, and other quality control methods specifically designed for laboratory rodents. Depending on the breadth of their activities, these laboratories most often include competence in serology, bacteriology, parasitology, and pathology. Virology and hematology expertise may also be required in some instances. Many larger research institutions have well-equipped and well-staffed institutional diagnostic laboratories. Testing services also can be obtained through commercial laboratories.

Traditionally, rodent diagnostic laboratories have tended to give highest priority to the investigation of clinical illnesses and necropsy evaluations of dead animals. That approach is no longer acceptable. While those services certainly are necessary, the needs of modern research and the principles of scientific method demand that diagnostic laboratories give greater priority to disease prevention. Most of the pathogen infections and pathogen-induced diseases of laboratory rodents are preventable.

REFERENCES

Bernard, C. 1865. An Introduction to the Study of Experimental Medicine (English translation by H. C. Greene, 1927). New York: MacMillan. 226 pp.

de Souza, M., and A. L. Smith. 1989. Comparison of isolation in cell culture with conventional and modified mouse antibody production tests for detection of murine viruses. J. Clin. Microbiol. 27:185-187.

DeLellis, R. A., and H. J. Wolfe. 1987. New techniques in gene product analysis. Arch. Pathol. Lab. Med. 111:620-627.

DiGiacomo, R. F., and T. D. Koepsell. 1986. Sampling for detection of infection or disease in animal populations. J. Am. Vet. Med. Assoc. 189:22-23.

Dubos, R. J., R. W. Schaedler, R. Costello, and P. Hoet. 1965. Indigenous, normal and autochthonous flora of the gastrointestinal tract. J. Exp. Med. 122:67-82.

Edberg, S. C. 1986. Nucleic acid hybridization analysis to elucidate microbial pathogens. Lab. Med. 17:735-738.

Hamm, T. E., Jr. 1983. The effects of health and health monitoring on oncology studies. Pp. 45-60 in The Importance of Laboratory Animal Genetics, Health and the Environment in Biomedical Research, E. C. Melby, Jr., and M. W. Balk, eds. New York: Academic Press.

Howanitz, J. H. 1988. Immunoassay. Arch. Pathol. Lab. Med. 112:771-779.

Hsu, C. K., A. E. New, and J. G. Mayo. 1980. Quality assurance of rodent models. Pp. 17-28 in Animal Quality and Models in Biomedical Research, A. Spiegel, S. Erichsen, and H. A. Solleveld, eds. Stuttgart: Gustav Fischer Verlag.

ILAR (Institute of Laboratory Animal Resources). 1976. Long-term holding of laboratory rodents. A report of the Committee on Long-Term Holding of Laboratory Rodents. ILAR News 19:L1-L25.

Iwai, H., T. Itoh, N. Nagiyama, and T. Nomura. 1980. Monitoring of murine infections in facilities for animal experimentation. Pp. 219-222 in Animal Quality and Models in Biomedical Research, A. Spiegel, S. Erichsen, and H. A. Solleveld, eds. Stuttgart: Gustav Fischer Verlag.

Jacoby, R. O., and S. W. Barthold. 1981. Quality assurance for rodents used in toxicological research and testing. Pp. 27-55 in Scientific Considerations in Monitoring and Evaluating Toxicological Research, E. J. Gralla, ed. Washington, D.C.: Hemisphere.

Kraft, V., and B. Meyer. 1986. Diagnosis of murine infections in relation to test method employed. Lab. Anim. Sci. 36:271-276.

Lindsey, J. R., D. B. Casebolt, and G. H. Cassell. 1986. Animal health in toxicological research: An appraisal of past performance and future prospects. Pp. 155-171 in Managing Conduct and Data Quality of Toxicology Studies. Princeton, N.J.: Princeton Scientific.

Loew, F. M., and J. G. Fox. 1983. Animal health surveillance and health delivery systems. Pp. 69-82 in The Mouse in Biomedical Research. Vol. III: Normative Biology, Immunology, and Husbandry, H. L. Foster, J. D. Small, and J. G. Fox, eds. New York: Academic Press.

NRC (National Research Council). 1985. Models for Biomedical Research: A New Perspective. A report of the Board on Basic Biology, Committee on Models for Biomedical Research. Washington, D.C.: National Academy Press. 180 pp.

NRC (National Research Council). 1991. Infectious Diseases of Mice and Rats. A report of the Institute of Laboratory Animal Resources Committee on Infectious Diseases of Mice and Rats. Washington, D.C.: National Academy Press. 397 pp.

Rowe, W. P., J. W. Hartley, J. D. Estes, and R. J. Huebner. 1959. Studies of mouse polyoma virus infection. I. Procedures for quantitation and detection of virus. J. Exp. Med. 109:379-391.

Rowe, W. P., J. W. Hartley, and R. J. Huebner. 1962. Polyoma and other indigenous mouse viruses. Pp. 131-142 in The Problems of Laboratory Animal Disease, R. J. C. Harris, ed. New York: Academic Press.

Savage, D. C. 1971. Defining the gastrointestinal microflora of laboratory mice. Pp. 60-73 in Defining the Laboratory Animal. Proceedings of the Fourth International Symposium on Laboratory Animals, organized by the International Committee on Laboratory Animals and the Institute of Laboratory Animal Resources and held April 8-11, 1969, in Washington, D.C. Washington, D.C.: National Academy of Sciences.

Sklar, J. 1985. DNA hybridization in diagnostic pathology. Human Pathol. 16:654-658.

Small, J. D. 1984. Rodent and lagomorph health surveillance–quality assurance. Pp. 709-723 in Laboratory Animal Medicine, J. G. Fox, B. J. Cohen, and F. M. Loew, eds. New York: Academic Press.

Smith, A. L. 1986a. Detection methods for rodent viruses. Pp. 123-142 in Complications of Viral and Mycoplasmal Infections in Rodents to Toxicology Research and Testing, T. E. Hamm, Jr., ed. Washington, D.C.: Hemisphere.

Smith, A. L. 1986b. Serologic tests for detection of antibody to rodent viruses. Pp. 731-751 in Viral and Mycoplosmal Infections of Laboratory Rodents: Effects on Biomedical Research, P. N. Bhatt, R. O. Jacoby, H. C. Morse III, and A. E. New, eds. Orlando, Fla.: Academic Press.

Smith, A. L. 1986c. Methods for potential application to rodent virus isolation and identification. Pp. 753-776 in Viral and Mycoplasmal Infections of Laboratory Rodents: Effects on Biomedical Research. P. N. Bhatt, R. O. Jacoby, H. C. Morse III, and A. E. New, eds. Orlando, Fla.: Academic Press.

Thigpen, J. E., and J. A. Tortorich. 1980. Recommended goals for microbiological quality control for laboratory animals used in the National Toxicology Program (NTP). Pp. 229-233 in Animal Quality and Models in Biomedical Research, A. Spiegel, S. Erichsen, and H. A. Solleveld, eds. Stuttgart: Gustav Fischer Verlag.

Van Der Logt, J. T. M. 1986. Serological study on the prevalence of murine viruses in laboratory animal colonies, in France and in the Netherlands (1981-1984). Sci. Tech. Anim. Lab.: 11:195-203.

PART II

Disease Agents

BACTERIA, FUNGI, AND VIRUSES

Adenoviruses

Agent. DNA virus. Two strains, MAd-1 and MAd-2 (formerly called FL and K87, respectively), are recognized.

Animals Affected. Mice and rats.

Epizootiology. Prevalence is probably low. MAd-1 is shed in urine, and MAd-2 is shed in feces. Transmission is by the oral route.

Clinical. Natural infection does not cause clinical disease.

Pathology. There are no pathologic lesions associated with natural infections of MAd-1. Viral inclusions in intestinal mucosa are associated with MAd-2 infections.

Diagnosis. The preferred diagnostic procedures are the ELISA and the IFA test, which test sera for antigens of both MAd-1 and MAd-2. Presumptive diagnosis of MAd-2 can be made by demonstration of characteristic intranuclear inclusions in histologic sections of intestinal epithelium. A fluorescent antibody method has been used for detecting MAd-2 antigen in the intestine. Definitive diagnosis requires virus isolation in tissue culture.

Control. Cesarean derivation and barrier maintenance may be necessary for eliminating either virus strain.

Interference with Research. MAd-1 can produce extensive persistent lesions in the kidneys of adult mice and render them more susceptible to experimental *Escherichia coli*-induced pyelonephritis. MAd (strain not given) has been reported to accelerate experimental scrapie in mice.

Suggested Reading

Hamelin, C., C. Jacques, and G. Lussier. 1988. Genome typing of mouse adenoviruses. J. Clin. Microbiol. 26:31-33.

Hamelin, C., and G. Lussier. 1988. Genotypic differences between the mouse adenovirus strains FL and K87. Experientia 44:65-66.

Luethans, T. N., and J. E. Wagner. 1983. A naturally occurring intestinal mouse adenovirus infection associated with negative serologic findings. Lab. Anim. Sci. 33:270-272.

Lussier G., A. L. Smith, D. Guenette, and J. P. Descoteaux. 1987. Serological relationship between mouse adenovirus strains FL and K87. Lab Anim. Sci. 37:55-57.

Otten, J. A., and R. W. Tennant. 1982. Mouse adenovirus. Pp. 335-340 in The Mouse in Biomedical Research. Vol. II: Diseases, H. L. Foster, J. D. Small, and J. G. Fox, eds. New York: Academic Press.

Smith, A. L., D. F. Winograd, and T. G. Burrage. 1986. Comparative biological characterization of mouse adenovirus strains FL and K87 and seroprevalence in laboratory rodents. Arch. Virol. 91:233-246.

Bacillus piliformis

Agent. Unclassified gram-negative bacterium, having vegetative and spore forms.

Animals Affected. Mice, rats, gerbils, hamsters, guinea pigs, rabbits, cats, dogs, nonhuman primates, horses, and others.

Epizootiology. Prevalence in the United States is unknown. Natural infection is thought to be caused by ingestion of spore-contaminated food or bedding.

Clinical. Subclinical infection is probably more common than clinical disease (commonly called Tyzzer's disease). Contributors to clinical disease include poor sanitation, overcrowding, transportation stress, food deprivation, dietary modifications, and altered host immune status. Clinical disease occurs most frequently in sucklings and weanlings, but animals of any age can be affected. Morbidity and mortality vary from low to high. Unexpected deaths, watery diarrhea, pasting of feces around the perineum, ruffled fur, and inactivity are the most common signs. Homozygous female and hemizygous male CBA/N-*xid* and C3.CBA/N-*xid* mice, which are deficient in a specific subpopulation of B cells, are highly susceptible, whereas T-cell-deficient athymic (*nu/nu*) mice are as resistant as immunocompetent mice.

Pathology. Primary infection occurs in the ileum and cecum, followed by ascension of the organisms by the portal vein to the liver and bacteremic spread to other tissues, most notably the myocardium. The organism preferentially replicates in the intestinal epithelium, smooth muscle, hepatocytes, and myocardium, but the degree of replication (and lesions) varies considerably between animals and between species. Intestinal lesions are usually more severe in rats than in mice. Myocardial lesions occur inconsistently. Gross lesions range from none to severe involvement of the intestine, liver, and/or heart. In mice, the most consistent finding is multiple pale to yellow foci in the liver. Infrequently, the ileum and cecum appear thickened, edematous, and hyperemic, and the myocardium contains circumscribed pale gray areas. Lesions in rats are similar, except that the ileum often appears dilated, atonic, and edematous (megaloileitis). The mesenteric lymph nodes are usually enlarged. In the ileum, in the cecum, and sometimes in the proximal colon, there is mild to severe loss of the mucosal epithelium, with blunting of villi in the ileum, thinning of the surface epithelium, and severe ulceration and hemorrhage. In more advanced stages, there is hyperplasia of crypt epithelium. Transmural acute to subacute inflammation can occur in areas of severe epithelial loss. In the liver there are multiple foci of coagulative necrosis that are rapidly converted to microabscesses. If the myocardium is affected, there is focal to diffuse myocardial necrosis with acute to subacute inflammation. In affected tissues, the characteristic large, filamentous bacilli are best demonstrated histologically in the cytoplasm of viable cells along the margin of necrotic tissues by the use of silver strains (Warthin-Starry or methenamine silver).

Diagnosis. Diagnosis of clinical disease is based on finding typical gross and microscopic lesions and characteristic organisms in silver-stained histologic sections. Both the IFA and the CF tests have been used for the diagnosis of subclinical infections, but neither test is available commercially in the United States. Alternatively, weanling animals can be immunosuppressed by the administration of 100-200 mg/kg cortisone acetate. Subclinical infection, if present, will become active disease, and characteristic lesions and organisms can be demonstrated histologically 7 days after cortisone administration. The presence of *B. piliformis* in tissues can be demonstrated by the finding of characteristic lesions and organisms histologically 5-7 days following inoculation of the tissue into gerbils or into homozygous *xid* female or hemizygous *xid* male mice.

Control. Cesarean-derivation and barrier-maintenance procedures, reduction of stress, and good sanitation procedures appear to minimize the occurrence of clinical disease. Good sanitation practices, avoidance of crowding, autoclaving of food and bedding, and the use of 0.3% sodium hypochlorite for disinfecting room surfaces are recommended for reducing spore contamination in conventional animal facilities. Oral administration of tetracycline can be helpful in controlling losses during outbreaks.

Interference with Research. Tyzzer's disease can cause high mortality in breeding colonies of mice and in mice used in long-term carcinogenesis studies. Administration of cortisone or adrenocorticotropic hormone, whole-body x-irradiation, transplantation of ascites tumors, and a high-protein diet can induce clinical disease. Tyzzer's disease alters the pharmacokinetics of warfarin and trimethoprim and the activity of hepatic transaminases.

Suggested Reading

Fries, A. S. 1980. Antibodies to *Bacillus piliformis* (Tyzzer's disease) in sera from man and other species. Pp. 249-250 in Animal Quality and Models in Research, A. Spiegel, S. Erichsen, and H. A. Solleveld, eds. Stuttgart: Gustav Fischer Verlag.

Fujiwara, K., M. Nakayama, and K. Takahashi. 1981. Serologic detection of inapparent Tyzzer's disease in rats. Jpn. J. Exp. Med. 51:197-200.

Ganaway, J. R. 1980. Effect of heat and selected chemical disinfectants upon infectivity of spores of *Bacillus piliformis* (Tyzzer's disease). Lab. Anim. Sci. 30:192-196.

Ganaway, J. R. 1982. Bacterial and mycotic diseases of the digestive system. Pp. 1-20 in The Mouse in Biomedical Research. Vol. II: Diseases, H. L. Foster, J. D. Small, and J. G. Fox, eds. New York: Academic Press.

Waggie, K. S., C. T. Hansen, J. R. Ganaway, and T. S. Spencer. 1981. A study of mouse strain susceptibility to *Bacillus piliformis* (Tyzzer's disease): The association of B-cell function and resistance. Lab. Anim. Sci. 31:139-142.

Weisbroth, S. H. 1979. Bacterial and mycotic diseases. Pp. 191-241 in The Laboratory Rat. Vol. 1: Biology and Diseases, H. J. Baker, J. R. Lindsey, and S. H. Weisbroth, eds. New York: Academic Press.

Cilia-Associated Respiratory Bacillus

Agent. Gram-negative bacterium.

Animals Affected. Laboratory rats and mice, wild rats (*Rattus norvegicus*), African white-tailed rats (*Mystromys albicaudatus*), and rabbits.

Epizootiology. Unknown.

Clinical. Clinical manifestations in rats are similar to those of severe murine respiratory mycoplasmosis (MRM) and can include hunched posture, ruffled coat, inactivity, head tilt, and accumulation of prophyrin pigment around the eyes and external nares. No description of clinical disease in mice has been published.

Pathology. In those instances in which cilia-associated respiratory (CAR) bacillus has been found in rats with natural disease, *Mycoplasma pulmonis* also was present. It is possible that *M. pulmonis* was the primary pathogen and the CAR bacillus increased disease severity. It is not known whether the CAR bacillus alone can cause natural clinical disease.

The predominant lesions in rats are those of advanced MRM due to *M. pulmonis* with some additional distinctive features. Severe bronchiectasis and bronchiolectasis, pulmonary abscesses, and atelectasis are associated with the accumulation of purulent or mucopurulent exudate in airways. An abundance of mucus often is present in peribronchiolar alveoli. Multifocal necrosis and acute inflammation of bronchiolar and bronchial epithelia often progress to severe granulomatous inflammation in airway walls and abscess formation in airway lumens. Disordered repair may result in distorted, scarred bronchioles and bronchiolitis obliterans. The ciliated border of the respiratory epithelium in affected airways often appears quite dense in hematoxylin- and eosin-stained sections because of the large numbers of CAR bacilli present between the cilia. The CAR bacillus can also be found on epithelial surfaces associated with MRM lesions in nasal passages, larynx, trachea, and middle ears. CAR bacillus-associated respiratory lesions similar to those seen in rats have been reported in C57BL/6J-*ob/ob* mice.

Diagnosis. Recognition of the argyrophilic CAR bacillus in Warthin-Starry silver-stained histologic sections of affected lungs has been the main diagnostic method used. The organism also can be demonstrated by transmission electron microscopy. The ELISA and the IFA test for this infection are in use in some laboratories but have not been fully evaluated.

Control. The infection probably can be eliminated by cesarean derivation, but definitive data are not available.

Interference with Research. Uncertain. The organism might be an important contributor to the morbidity and mortality caused by MRM in rats.

Suggested Reading

Ganaway, J. R., T. H. Spencer, T. D. Moore, and A. M. Allen. 1985. Isolation, propagation, and characterization of a newly recognized pathogen, cilia-associated respiratory (CAR) bacillus of rats, an etiological agent of chronic respiratory disease. Infect. Immun. 47:472-479.

Griffith, J. W., W. J. White, P. J. Danneman, and C. M. Lang. 1988. Cilia-associated respiratory bacillus infection of obese mice. Vet. Pathol. 25:72-76.

Mackenzie, W. F., L. S. Magill, and M. Hulse. 1981. A filamentous bacterium associated with respiratory disease in wild rats. Vet. Pathol. 18:836-839.

Matsushita, S. 1986. Spontaneous respiratory disease associated with cilia-associated respiratory (CAR) bacillus in a rat. Jpn. J. Vet. Sci. 48:437-440.

Matsushita, S., M. Kashima, and H. Joshima. 1987. Serodiagnosis of cilia-associated respiratory bacillus infection by the indirect immunofluorescence assay technique. Lab. Anim. (London) 21:356-359.

van Zwieten, M. J., H. A. Solleveld, J. R. Lindsey, F. G. deGroot, C. Zurcher, and C. F. Hollander. 1980. Respiratory disease in rats associated with a filamentous bacterium. Lab. Anim. Sci. 30:215-221.

Citrobacter freundii Biotype 4280

Agent. Gram-negative bacterium. Usually considered an opportunistic pathogen.

Animals Affected. Mice.

Epizootiology. Transmission is presumed to be by the fecal-oral route. The organism is rarely found in cesarean-derived, barrier-maintained mice.

Clinical. Signs of disease are nonspecific and include ruffled fur, listlessness, weight loss, stunting, pasty feces around the anus and perineum, and rectal prolapse. Suckling mice are more susceptible to disease than are adults. Mortality can reach 60%, and the occurrence of rectal prolapse can reach 15%. Mortality is significantly higher in C3H/HeJ than in DBA/2J, C57BL/6J, or N:NIHS (Swiss) mice.

Pathology. Infection in mice lasts only about 4 weeks. Even if the infection is eliminated as early as 2 days postinfection by administration of neomycin sulfate or tetracycline hydrochloride, mucosal hyperplasia still occurs. Presence of the infection in the intestine for 10 days results in maximum hyperplasia. The descending colon is most commonly affected, but the entire colon and cecum can be involved. Grossly, affected bowel is thickened and rigid in appearance. Microscopically, crypt height is increased threefold, mitotic activity is increased, goblet cells are decreased, and basophilia of the epithelium is increased. Crypt abscesses are common, and mucosal erosions and ulcers can occur. The occurrence of necrotizing and inflammatory lesions tends to parallel mortality. Variable numbers of neutrophils or mononuclear leukocytes can be present in the lamina propria, but there is often a paucity of inflammatory cells. Goblet-cell hyperplasia with mucinous distension of crypts and streaming of mucin into the gut lumen can occur during regression of mucosal hyperplasia.

Diagnosis. Diagnosis is by demonstration of characteristic lesions in the large intestine and isolation of organisms of the pathogenic biotype.

Control. Definitive data are lacking. Control probably requires depopulation and restocking with cesarean-derived mice. Neomycin or tetracycline administered in drinking water reduces losses during outbreaks but probably does not completely eliminate infection.

Interference with Research. The cytokinetics of the mucosal epithelium in the large intestine is profoundly altered in infected mice. Susceptibility to the carcinogen 1,2-dimethylhydrazine is increased, and the latent period for neoplasia induction is reduced.

Suggested Reading

Barthold, S. W. 1980. The microbiology of transmissible murine colonic hyperplasia. Lab. Anim. Sci. 30:167-173.

Barthold, S. W., G. L. Coleman, P. N. Bhatt, G. W. Osbaldiston, and A. M. Jonas. 1976. The etiology of transmissible murine colonic hyperplasia. Lab. Anim. Sci. 26:889-894.

Barthold, S. W., G. L. Coleman, R. O. Jacoby, E. M. Livstone, and A. M. Jonas. 1978. Transmissible murine colonic hyperplasia. Vet. Pathol. 15:223-236.

Johnson, E., and S. W. Barthold. 1979. The ultrastructure of transmissible murine colonic hyperplasia. Am. J. Pathol. 97:291-313.

Corynebacterium kutscheri

Agent. Gram-positive bacterium.

Animals Affected. Mice, rats, and rarely guinea pigs.

Epizootiology. Persistent subclinical infections are thought to be common in conventionally reared stocks and rare in barrier-maintained stocks. The main sites of infection are probably the oropharynx, submaxillary lymph nodes, and large intestine, with transmission mainly by the fecal-oral route.

Clinical Signs. Infection is usually inapparent. Disease occurs only after the immune system is compromised by experimental procedures, dietary deficiency, or concurrent infections with other agents. Signs of clinical disease in rats are usually those of a respiratory infection: dyspnea, rales, weight loss, humped posture, and anorexia. Signs in mice are usually those of severe septicemia: dead and moribund animals. Arthritis or abscesses can occur in either species.

Pathology. Septicemia results in septic emboli in many organs. In mice, large bacterial emboli lodge in capillary beds, particularly in the kidney and liver. Embolic glomerulitis is characteristic. Abscess formation can occur at the focus of infection. In rats, bacterial emboli lodge in the capillaries of the lungs. Alveoli become packed with polymorphonuclear leukocytes and can form large necropurulent centers. Fibrinous or fibrous pleuritis often develops. Occasionally, abscesses occur in the liver, kidneys, subcutis, peritoneal cavity, and other sites.

Diagnosis. Detection of persistent subclinical infections is very difficult. Available methods are unsatisfactory for detecting persistent subclinical infections, although an ELISA has been developed that shows promise for detecting antibodies. There is also a new DNA probe method, but its usefulness has not been determined. Cortisone acetate can be given to provoke active disease, which is then diagnosed by culture of the organism, demonstration of characteristic lesions, and exclusion of other infectious agents and disease processes.

Control. Cesarean derivation and barrier maintenance are effective means of control.

Interference with Research. Infection with *C. kutscheri* can complicate experiments involving immunologically compromised mice or rats.

Suggested Reading

Ackerman, J. I., J. G. Fox, and J. C. Murphy. 1984. An enzyme linked immunosorbent assay for detection of antibodies to *Corynebacterium kutscheri*. Lab. Anim. Sci. 34:38-43.

Barthold, S. W., and D. G. Brownstein. 1988. The effect of selected viruses on *Corynebacterium kutscheri* infection in rats. Lab. Anim. Sci. 38:580-583.

Brownstein, D. G., S. W. Barthold, R. L. Adams, G. A. Terwilliger, and J. G. Aftosmis. 1985. Experimental *Corynebacterium kutscheri* infection in rats: Bacteriology and serology. Lab. Anim. Sci. 36:135-138.

Saltzgaber-Muller, J., and B. A. Stone. 1986. Detection of *Corynebacterium kutscheri* in animal tissues by DNA-DNA hybridization. J. Clin. Microbiol. 24:759-763.

Suzuki, E., K. Mochida, and M. Nakagawa. 1988. Naturally occurring subclinical *Corynebacterium kutscheri* infection in laboratory rats: Strain and age related antibody response. Lab. Anim. Sci. 38:42-45.

Weisbroth, S. H., and S. Scher. 1968. *Corynebacterium kutscheri* infection in the mouse. II. Diagnostic serology. Lab. Anim. Care 18:459-468.

Cytomegalovirus, Mouse

Agent. Double-stranded DNA virus, family Herpesviridae.

Animals Affected. Wild mice and laboratory mice that have contracted the infection from wild mice.

Epizootiology. Prevalence in laboratory mice is uncertain but is probably very low. Transmission occurs through the saliva, and infection persists throughout life.

Vertical transmission also can occur but has not been fully explained. Direct passage of the virus across the placenta to the fetus and/or transmission via germ cells have been suggested.

Clinical Signs. Natural infections are subclinical.

Pathology. In natural infections, large acidophilic intranuclear inclusions are found in salivary gland acinar and duct cells. Affected cells typically are enlarged three to four times normal. The submaxillary glands are affected most, the sublingual glands less, and the parotid glands least.

Diagnosis. The IFA and CF tests have been shown to be sensitive for acute experimental infections, and the ELISA has been shown to be sensitive for persistent experimental infections. These methods may prove useful for monitoring laboratory mice for cytomegalovirus infection in selected situations. Virus isolation can be accomplished using mouse embryo fibroblasts or other tissue culture systems.

Control. Wild mice must be excluded from rodent facilities.

Interference with Research. Natural infections have not been reported to interfere with research.

Suggested Reading

Classen, D. C., J. M. Moringstar, and J. D. Shanley. 1987. Detection of antibody to murine cytomegalovirus by enzyme-linked immunosorbent and indirect immunofluorescence assays. J. Clin. Microbiol. 25:600-604.

Lussier, G., D. Guenette, and J. P. Descoteaux. 1987. Comparison of serological tests for the detection of antibody to natural and experimental murine cytomegalovirus. Can. J. Vet. Res. 51:249-252.

Mercer, J. A., C. A. Wiley, and D. H. Spector. 1988. Pathogenesis of murine cytomegalovrirus infection: Identification of infected cells in the spleen during acute and latent infections. J. Virol. 62:987-997.

Mims, C. A., and J. Gould. 1979. Infection of salivary glands, kidneys, adrenals, ovaries, and epithelia by murine cytomegalovirus. J. Med. Microbiol. 12:113-122.

Osborn, J. E. 1982. Cytomegalovirus and other herpesviruses. Pp. 267-292 in The Mouse in Biomedical Research. Vol. II: Diseases, H. L. Foster, J. D. Small, and J. G. Fox, eds. New York: Academic Press.

Quinnan, G. V., and J. E. Manischewitz. 1987. Genetically determined resistance to lethal murine cytomegalovirus infection is mediated by interferon-dependent and -independent restriction of virus replication. J. Virol. 61:1875-1881.

Ectromelia Virus

Agent. DNA virus, family Poxviridae.

Animals Affected. Mice.

Epizootiology. Ectromelia virus is possibly enzootic in some mouse colonies. Periodic epizootics have occurred in the United States since 1950, most commonly in research laboratories that exchange live mice and their tissues, sera, or transplantable tumors. Natural transmission is dependent on direct contact and fomites. Skin abrasions are thought to be the main route of virus entry. Infected animals begin shedding virus about 10 days postinfection when characteristic skin lesions appear. Persistent infection (carrier state) was previously thought to be important in the epizootiology of mousepox. More recent data suggest that significant numbers of virus particles are shed from skin lesions for only about 3 weeks.

Clinical. The severity of clinical disease (mousepox) varies greatly and depends on mouse strain, virus strain, length of time infection has been present in the colony, and husbandry practices. Inapparent infections occur mainly in resistant inbred mouse strains such as C57BL/6 or C57BL/10. Resistance in the these strains appears to be due to a single autosomal dominant trait. The more susceptible mouse strains include A, CBA, C3H, DBA/2, and BALB/c. Clinical manifestations include one or more of the following: ruffled hair; hunched posture; facial edema; conjunctivitis; swelling of the feet; cutaneous papules, erosions, or encrustations, mainly on the face, ears, feet, or tail; and necrotic amputation (ectromelia) of limbs or tails. Mortality varies from less than 1% to greater than 80%.

Pathology. The incubation period is 7-10 days. Virus replicates in the skin and then in the regional lymph nodes, resulting in a mild primary viremia. Virus is taken up by splenic and hepatic macrophages, where there is extensive multiplication that results in a massive secondary viremia and sometimes in death due to diffuse splenic and hepatic necrosis. Virus from the secondary viremia localizes in a wide variety of tissues, especially the skin (basal cells), conjunctiva, and lymphoid tissues. A primary lesion may appear at the site of skin inoculation about 4-7 days postinfection. Foot swelling and secondary generalized rash (pocks) may appear 7-10 days postinfection. Skin lesions heal within 2 weeks, leaving scars. In acute mousepox, there is severe necrosis of liver, spleen, lymph nodes, Peyer's patches, and thymus. Jejunal hemorrhage often results from mucosal erosions. Characteristic large eosinophilic cytoplasmic inclusions may be present in skin lesions.

Diagnosis. The ELISA is sensitive and specific in unvaccinated mice; however, it may give false-positive results in NZW and NZB mice. The HAI is relatively insensitive but does not give positive reactions to sera from mice vaccinated with the IHD-T strain of vaccinia virus. Diagnosis of acute disease is based on serologic testing of survivors or on the demonstration, using transmission electron microscopy, of characteristic large virus particles in affected tissues. Differential diagnosis of skin lesions should exclude bite wounds and loss of limbs due to *Streptobacillus moniliformis*. Biologic materials such as cells and blood can be screened for ectromelia virus by injecting the tissue into known pathogen-free mice followed by serologic testing.

Control. Quarantine and testing of incoming mice and mouse tissues from sources other than commercial barrier facilities are the best way to prevent the introduction of infection. In the past, the accepted practice for eradicating ectromelia virus was elimination of infected mouse colonies and all infected biologic materials, along with rigorous decontamination of rooms and equipment. Cesarean derivation of infected mouse stocks was not acceptable because intrauterine infection is known to occur in mice infected during pregnancy. More recently it has been suggested that quarantine and cessation of breeding might successfully eliminate the virus. Vaccination with a live-virus vaccine, the IHD-T strain of vaccinia virus adapted to growth in embryonated eggs, can be useful in eliminating disease from small closed colonies where all offspring can be vaccinated by 6 weeks of age. Vaccination can protect mice from fatal disease but does not prevent infection or virus transmission.

Interference with Research. Up to 100% of the animals in an experiment can die in an explosive outbreak. Manipulations that exacerbate ectromelia virus infections or promote epizootics include experimental infection with tubercle bacilli, x-irradiation, administration of various toxic chemicals, shipping, tissue transplantation, castration, and tumors. Ectromelia virus infection can alter phagocytic response. Conversely, procedures that decrease phagocytosis can increase susceptibility to ectromelia virus, e.g., large doses of endotoxin or splenectomy.

Suggested Reading

AALAS (American Association of Laboratory Animal Science). 1981. Ectromelia (mousepox) in the United States. Proceedings of a seminar presented at the 31st Annual Meeting of AALAS held in Indianapolis, Indiana, October 8, 1980. Lab. Anim. Sci. 31:549-631.

Bhatt, P. N., and R. O. Jacoby. 1987. Mousepox in inbred mice innately resistant or susceptible to lethal infection with ectromelia virus. I. Clinical responses. Lab. Anim. Sci. 37:11-15.

Bhatt, P. N., and R. O. Jacoby. 1987. Mousepox in inbred mice innately resistant or susceptible to lethal infection with ectromelia virus. III. Experimental transmission of infection and derivation of virus-free progeny from previously infected dams. Lab. Anim. Sci. 37:23-28.

Bhatt, P. N., and R. O. Jacoby. 1987. Effect of vaccination on the clinical response, pathogenesis and transmission of mousepox. Lab. Anim. Sci. 37:610-614.

Bhatt, P. N., R. O. Jacoby, and L. Gras. 1988. Mousepox in inbred mice innately resistant or susceptible to lethal infection with ectromelia virus. IV. Studies with the Moscow strain. Arch. Virol. 100:221-230.

Fenner, F. 1982. Mousepox. Pp. 209-230 in The Mouse in Biomedical Research. Vol. II: Diseases, H. L. Foster, J. D. Small, and J. G. Fox, eds. New York: Academic Press.

Encephalitozoon cuniculi

Agent. A protozoan, order Microsporidia.

Animals Affected. Mice, rats, rabbits, hamsters, guinea pigs, humans, and many other mammals.

Epizootiology. Prevalence in mouse and rat stocks is not known but is thought to be very low in comparison with that in rabbits, in which the organism is considered ubiquitous. Rabbits undoubtedly provide the major source of infection for mice and rats in research facilities. Spores of *E. cuniculi* are shed in the urine and ingested by another host.

Clinical. Natural infections usually are inapparent.

Pathology. In cases of clinical disease, the classic lesion in rats and rabbits is meningoencephalitis with multifocal granulomatous inflammation. Activated macrophages form glial nodules in response to the organism. These nodules have necrotic centers or appear as solid sheets of cells. Varying numbers of lymphocytes and plasma cells are seen in the meninges and around vessels. The brain lesions in mice are similar, except for the lack of the granulomatous foci. The organism occurs intracellularly in the renal tubular epithelium with or without the presence of an inflammatory response. In chronic infections, focal destruction of tubules and replacement by fibrous connective tissue results in small pits on the cortical surface. Lesions in organs other than the kidney and brain are less consistent. Intraperitoneal inoculation of *E. cuniculi*, as when contaminated transplantable tumors are passaged, results in ascites in mice.

Diagnosis. Several serologic tests have been developed for diagnosis of the infection in rabbits; however, only the IFA and immunoperoxidase tests have been used in surveying mouse and rat colonies. Other methods used for diagnosis include detection of parasites in urine, demonstration of typical lesions and organisms in tissue sections, and an intradermal skin test.

Control. Mice and rats should not be exposed to infected rabbits. Serologic testing of adult animals with selection of *E. cuniculi*-free breeding stocks has been used successfully for eradicating the infection in rabbits.

Interference with Research. The histologic changes caused by *E. cuniculi* infection in the brain and kidneys can complicate the interpretation of lesions in studies requiring histopathology. *E. cuniculi* can contaminate transplantable tumors and alter host responses during tumor passage in mice. Mice experimentally infected with *E. cuniculi* have reduced humoral antibody titers to sheep erythrocytes, reduced proliferative spleen cell responses to mitogens, and altered natural killer cell activity.

Suggested Reading

Beckwith, C., N. Peterson, J. J. Liu. and J. A. Shadduck. 1988. Dot enzyme-linked immunosorbent assay (dot ELISA) for antibodies to *Encephalitozoon cuniculi*. Lab. Anim. Sci. 38:573-576.

Didier, E. S., and J. A. Shadduck. 1988. Modulated immune responsiveness associated with experimental *Encephalitozoon cuniculi* infection in BALB/c mice. Lab. Anim. Sci. 38:680-684.

Gannon, J. 1980. The course of infection of *Encephalitozoon cuniculi* in immunodeficient and immunocompetent mice. Lab. Anim. (London) 14:189-192.

Majeed, S. K., and A. J. Zubaidy. 1982. Histopathological lesions associated with *Encephalitozoon cuniculi* (nosematosis) infection in a colony of Wistar rats. Lab. Anim. (London) 16:244-247.

Niederkorn, J. Y., J. A. Shadduck, and E. C. Schmidt. 1981. Susceptibility of selected inbred strains of mice to *Encephalitozoon cuniculi*. J. Infect. Dis. 144:249-253.

Shadduck, J. A., and S. P. Pakes. 1971. Encephalitozoonosis (nosematosis) and toxoplasmosis. Am. J. Pathol. 64:657-674.

Hantaviruses

Agent. Single-stranded RNA viruses, family Bunyaviridae, genus *Hantavirus*. Hantaan virus, the cause of Korean hemorrhagic fever in humans, is the prototype member of the genus and is a significant zoonotic pathogen of laboratory rodents.

Animals Affected. Natural hosts of all hantaviruses appear to be small mammals, primarily rodents. Multiple species may serve as hosts in a given geographical area, and strain of virus and likelihood of causing disease in humans vary from region to region. Thus far, about five dominant associations between hantaviruses, rodent carriers, and human diseases (if present) have been described as follows: Hantaan virus, the field mouse *Apodemus agrarius*, Korean hemorrhagic fever (KHF) in Korea, and the severe form of epidemic hemorrhagic fever in China; Puumala virus, the bank vole *Clethrionomys glareolus*, nephropathia epidemica (NE) in Eastern Europe and Scandinavia; urban and laboratory rat viruses, *Rattus norvegicus*, moderate disease found mostly in people in Asia but occasionally also in Europeans; Girard Point and other viruses from North and South America, *Rattus norvegicus*, no disease recognized in humans, although serologic evidence of infection has been found; and Prospect Hill virus, the meadow vole *Microtus pennsylvanicus*, no disease recognized in humans. Naturally infected laboratory rats have been the source of *Hantavirus* infections in research personnel in Japan, Belgium, the United Kingdom, and France.

Epizootiology. Hantaan virus appears in the lungs of its reservoir host about 10 days postinfection and subsequently appears in the urine and saliva. Peak virus shedding occurs about 3 weeks after infection, but virus can be detected in the lungs for 6 months and occasionally for up to 2 years. Aerosols are the main mode of transmission. Other hantaviruses are assumed to have similar patterns of infection in their reservoir hosts.

Hantaviruses are transmitted to humans from persistently infected rodents and other small mammals. In laboratory settings, this is usually from laboratory rats or their tumors. The major mode of transmission is by aerosols of urine, feces, or saliva containing infectious virus. Direct animal contact is not necessary. Animal bites can transmit the infection but appear to be of relatively minor importance.

Clinical. Reservoir hosts have persistent subclinical infections. Human disease has occurred only in Europe and Asia.*

---

*Signs of KHF in humans vary from mild to severe and include fever, headache, muscular pains, hemorrhages (cutaneous petechia or ecchymoses, hemoptysis, hematuria, hematemesis, melena), and proteinuria.

Pathology. Lesions have not been observed in reservoir hosts infected with hantaviruses. Lungs and other tissues contain large amounts of virus without morphologic changes.

Diagnosis. Since *Hantavirus* infections in rodents are inapparent, diagnosis is most likely to be made through health surveillance. Recommended serologic tests are the IFA test, the ELISA, and the HAI test. Noninfectious antigen should be used, and work with animals and blood products should be done in a biological safety cabinet. P3 conditions are needed for working with unconcentrated virus in small amounts. The RAP test is recommended for testing transplantable tumors and other biologic materials for *Hantavirus* contamination.

Control. *Hantavirus* infections can be prevented by obtaining animals, transplantable tumors, and other biologic materials that have been tested and found to be free of infection. Contamination of laboratory rodent stocks by wild rodents must be prevented.

Laboratory rodent stocks found to be infected with a *Hantavirus* should be destroyed and replaced with pathogen-free stock. Although not proven to be effective, cesarean derivation has been recommended for eliminating the infection in valuable genetic stocks.

Interference with Research. Some hantaviruses are important zoonoses.

Suggested Reading

Childs, J. E., G. E. Glass, G. W. Korch, J. W. LeDuc. 1987. Prospective seroepidemiology of hantaviruses and population dynamics of small mammal communities of Baltimore, Maryland. Am. J. Trop. Med. Hyg. 37:648-662.

Hemorrhagic Fever with Renal Syndrome (HFRS) Virus Guideline Committee. 1986. Guidelines for surveillance, prevention, and control of Hantaan virus infection in laboratory animal colonies. Pp. 209-216 in Viral and Mycoplasmal Infections of Laboratory Rodents: Effects on Biomedical Research, P. N. Bhatt, R. O. Jacoby, H. C. Morse III, and A. E. New, eds. Orlando, Fla.: Academic Press.

Johnson, K. 1986. Hemorrhagic fever—Hantaan virus. Pp. 193-207 in Viral and Mycoplasmal Infections of Laboratory Rodents: Effects on Biomedical Research, P. N. Bhatt, R. O. Jacoby, H. C. Morse III, and A. E. New, eds. Orlando, Fla.: Academic Press.

Kawamata, J., T. Yamanouchi, K. Dohmae, H. Miyamoto, M. Takahaski, K. Yamaniski, T. Kurata, and H. W. Lee. 1987. Control of laboratory acquired hemorrhagic fever with renal syndrome (HFRS) in Japan. Lab. Anim. Sci. 37:431-436.

Tsai, T. F. 1987. Hemorrhagic fever with renal syndrome: Mode of transmission to humans. Lab. Anim. Sci. 37:428-430.

Tsai, T. F. 1987. Hemorrhagic fever with renal syndrome: Clinical aspects. Lab. Anim. Sci. 37:419-427.

Hepatitis Virus, Mouse

Agent. Single-stranded RNA virus, family Coronaviridae, genus *Coronavirus*.
Animals Affected. Mice.

Epizootiology. Mouse hepatitis virus (MHV) infection is extremely contagious; prevalence can exceed 80%. Pathogenesis is influenced by such factors as virus strain and mouse strain. Current evidence suggests that the infection runs its course in 2-3 weeks, and there is no carrier state. Transmission is by direct contact, fomites, and airborne particles. MHV is a frequent contaminant of transplantable tumors and cell lines.

Clinical. In immunocompetent mice, MHV infections are usually subclinical. Infant mice of naive breeding populations can show diarrhea and high mortality when infected with the more virulent enterotropic MHV strains. Athymic (*nu/nu*) mice show progressive emaciation leading to debility and death.

Pathology. Strains of MHV differ greatly in virulence and tissue tropism, and mouse strains differ greatly in susceptibility to MHV. These factors interact with host age and route and dose of virus inoculation to determine the outcome of infection. Mechanisms of host resistance to MHV infection are poorly understood. Mice are fully susceptible to the virus as neonates, but some strains acquire resistance at 2-3 weeks of age as lymphoreticular function matures. Cell-mediated immunity is important in the development of resistance; humoral immunity is considered relatively unimportant. Macrophages, interferon, and natural killer cells may also have important roles.

There are two major disease patterns: the respiratory pattern and the enteric pattern. Lesions in immunocompetent mice are present for only about 7-10 days and are usually nonspecific and subtle, particularly those associated with the respiratory pattern. In the respiratory pattern, infection involves the nasal passages and lungs; intestinal involvement is minimal. Lesions include mild olfactory mucosal necrosis, neuronal necrosis of olfactory bulbs and tracts, lymphoplasmacytic infiltrates and vacuolation in the brain, multifocal interstitial pneumonia with mild perivascular lymphoid infiltrates, and multifocal necrotizing hepatitis. In the enteric pattern, infection is primarily restricted to the bowel, with variable spread to other abdominal organs such as the liver and abdominal lymph nodes. Lesions are most severe in neonatal mice because of their relatively slow kinetics of mucosal epithelium turnover. Varying degrees of epithelial lysis and blunting of villi occur in the small intestine. Numerous multinucleate syncytial giant cells (balloon cells) can occur on the villi, as well as in the crypts. In more severe cases, there can be ulceration of the mucosa. A similar lytic process occurs in the ascending colon and cecum. Occasionally, there is multifocal necrotizing hepatitis and/or encephalitis.

MHV infection in athymic (*nu/nu*) or neonatally thymectomized mice becomes progressively more generalized, severe, and chronic, with involvement of many organs, including the liver, intestine, lungs, bone marrow, lymphoreticular organs, vascular endothelium, and brain. Multifocal necrosis with syncytial giant cells usually occurs in the liver. Splenomegaly may develop because of compensatory myelopoiesis, and large numbers of myelopoietic cells may appear in the liver.

Diagnosis. The ELISA is the test of choice for serologic monitoring. An IFA test is also available and is about equal in sensitivity to the ELISA. CF and serum neutralization tests are less sensitive. Heterozygotes or sentinel mice should be used to test athymic stocks, because *nu/nu* homozygotes do not develop CF antibody to MHV and have a weak and variable antibody response to the ELISA and the serum neutralization test. The characteristic histologic lesions of MHV infections are useful in both health surveillance and necropsy diagnosis. Virus isolation can be difficult because not all strains grow equally well in all cell lines. Transplantable tumors and other biologic materials can be screened by virus isolation or by the MAP test.

Control. Strict adherence to barrier protocol, regular health surveillance, and testing of biologic materials from mice are necessary to prevent MHV infection. Cesarean derivation and barrier maintenance traditionally have been recommended for rederivation of breeding stocks; however, these measures may be unnecessary because recent evidence suggests that the virus is present for only 2-3 weeks. Another approach is to isolate individual breeding pairs from MHV-infected populations in separate containment devices, such as filter-top cage systems, and subsequently to select seronegative progeny as breeders.

Interference with Research. MHV has been reported to alter many experimental results. Examples are alterations in immune function and hepatic enzyme activities; inhibition of lymphocyte proliferative responses in mixed lymphocyte cultures and mitogen-stimulated cells; alteration of phagocytic and tumorcidal activity; increase of hepatic uptake of injected iron; increase of susceptibility to other indigenous pathogens; activation of natural killer cells and production of interferon; delay of the increase in plasma lactic dehydrogenase activity following infection with lactic dehydrogenase virus; and occurrence of anemia, leukopenia, and thrombocytopenia. In athymic (*nu/nu*) mice, the virus can also cause spontaneous differentiation of lymphocytes bearing T-cell markers, alter IgM and IgG responses to sheep erythrocytes, enhance phagocytic activity of macrophages, cause rejection of xenograft tumors, impair liver regeneration after partial hepatectomy, and cause hepatosplenic myelopoiesis. Subclinical infections are exacerbated by thymectomy; whole-body irradiation; reticuloendothelial blockade by iron salts; and administration of cortisone, cyclophosphamide, antilymphocyte serum, chemotherapeutic agents, or halothane anesthesia.

Suggested Reading

Barthold, S. W. 1986. Research complications and state of knowledge of rodent coronaviruses. Pp. 53-89 in Complications of Viral and Mycoplasma Infections in Rodents to Toxicology Research and Testing, T. E. Hamm, Jr., ed. Washington, D.C.: Hemisphere.

Barthold, S. W. 1986. Mouse hepatitis virus biology and epizootiology. Pp. 571-601 in Viral and Mycoplasmal Infections of Laboratory Rodents: Effects on Biomedical Research, P. N. Bhatt, R. O. Jacoby, H. C. Morse III, and A. E. New, eds. Orlando, Fla.: Academic Press.

Barthold, S. W. 1987. Host age and genotype effects on enterotropic mouse hepatitis virus infection. Lab. Anim. Sci. 37:36-40.

Barthold, S. W., and A. L. Smith. 1984. Mouse hepatitis virus strain-related patterns of tissue tropism in suckling mice. Arch. Virol. 81:103-112.

Boyle, J. F., D. G. Weismiller, and K. V. Holmes. 1987. Genetic resistance to mouse hepatitis virus correlates with absence of virus-binding activity on target tissues. J. Virol. 61:185-189.

Holmes, K. V., J. F. Boyle, and M. F. Frana. 1986. Mouse hepatitis virus: Molecular biology and implications for pathogenesis. Pp. 603-624 in Viral and Mycoplasmal Infections of Laboratory Rodents: Effects on Biomedical Research, P. N. Bhatt, R. O. Jacoby, H. C. Morse III, and A. E. New, eds. Orlando, Fla.: Academic Press.

H-1 Virus*

Agent. Single-stranded DNA virus, family Parvoviridae, genus *Parvovirus*.

Animals Affected. Laboratory and wild rats (*Rattus norvegicus*).

Epizootiology. Common infection of wild and laboratory rats; prevalance exceeds 50% in some populations. Epizootiological characteristics are generally assumed to be similar to those of Kilham rat virus infection. Transmission is primarily horizontal, and virus is shed in urine, feces, nasal secretions, and milk. Transplacental infection has not been demonstrated in natural infections.

Clinical. Natural infections are inapparent.

Pathology. There are no pathologic changes associated with natural infections.

Diagnosis. The ELISA and IFA test are usually used for initial screening, followed by either the HAI, CF, or neutralization test for discriminating between H-1 virus and Kilham rat virus infections. Primary rat embryo, 324K, or BHK-21 cells can be used for virus isolation.

Control. The same measures recommended for Kilham rat virus (see page 25) should be effective in controlling H-1 virus.

Interference with Research. It might be possible for H-1 virus to alter studies of fetal development or teratogenesis, but this has not been reported to occur as a result of natural infection. H-1 virus has been reported to cause hepatocellular necrosis when rats are subjected to liver injury by hepatotoxic chemicals, parasitism, or partial hepatectomy. H-1 virus has been reported to inhibit experimental tumor induction by adenovirus 12 and dimethylbenzanthracene in hamsters.

Suggested Reading

Jacoby, R. O., P. N. Bhatt, and A. M. Jonas. 1979. Viral diseases. Pp. 273-283 in The Laboratory Rat. Vol. I: Biology and Diseases, H. J. Baker, J. R. Lindsey, and S. H. Weisbroth, eds. New York: Academic Press.

Kilham, L. 1966. Viruses of laboratory and wild rats. Pp. 117-140 in Viruses of Laboratory

---

*This virus was designated "H-1" by Toolan in 1961, and this is its official taxonomic name. In the United States, the virus is commonly called Toolan H-1 virus.

Rodents, R. Holdenried, ed. National Cancer Institute Monograph 20. Washington, D.C.: U.S. Department of Health, Education, and Welfare.
Ruffolo, P. R., G. Margolis, and L. Kilham. 1966. The induction of hepatitis by prior partial hepatectomy in resistant adult rats infected with H-1 virus. Am. J. Pathol. 49:795-824.
Tattersall, P., and S. F. Cotmore. 1986. The rodent parvoviruses. Pp. 305-348 in Viral and Mycoplasmal Infections of Laboratory Rodents: Effects on Biomedical Research, P. N. Bhatt, R. O. Jacoby, H. C. Morse III, and A. E. New, eds. Orlando, Fla.: Academic Press.
Toolan, H. W., S. L. Rhode III, and J. F. Gierthy. 1982. Inhibition of 7,12-dimethylbenz(a)anthracene tumors in Syrian hamsters by prior infection with H-1 parvoviruses. Cancer Res. 42:2552-2555.

Kilham Rat Virus

Agents. Single-stranded DNA virus, family Parvoviridae, genus *Parvovirus*.
Animals Affected. Laboratory and wild rats are the natural hosts.
Epizootiology. Kilham rat virus (KRV) is a common infection in wild and laboratory rats; prevalence exceeds 50% in some populations. Transmission is primarily by the horizontal route, either through direct contact or fomites. Virus is shed in urine, feces, milk, and nasal secretions. Transplacental infection is not considered important. Persistent infection can last up to 14 weeks. KRV is a frequent contaminant of cultured cell lines and transplantable tumors.
Clinical. KRV infections rarely cause clinical disease. Signs in the few instances of overt disease that have been reported include increased numbers of uterine resorption sites in pregnant dams and runting, ataxia, cerebellar hypoplasia, and jaundice in their pups. Spontaneous deaths, cerebellar hypoplasia, scrotal cyanosis, jaundice, abdominal swelling, dehydration, and other signs of severe illness have occurred in juvenile and young adult rats.
Pathology. Parvoviruses attack rapidly dividing cells. In newborn and young rats, KRV can cause jaundice, hemorrhagic infarction with thrombosis in multiple organs (including brain, spinal cord, testes, and epididymis), and cerebellar hypoplasia. Amphophilic intranuclear inclusions occur in the endothelium and other cells of affected organs. Focal necrosis, hypertrophy and vacuolar degeneration of hepatocytes, cholangitis, and biliary hyperplasia also occur. Hemorrhagic encephalopathy has been reported in naturally infected LEW rats given cyclophosphamide.
Diagnosis. The ELISA and IFA test are the most sensitive tests but do not discriminate between different serotypes of parvoviruses. The HAI, CF, and neutralization tests are used for serotype discrimination. In clinical disease, typical lesions should be demonstrated, and virus isolation should be carried out. The virus can be grown in primary rat embryo, 324K, and BHK-21 cells.
Control. The most practical approach to controlling infection is to obtain animals demonstrated free of KRV by serologic monitoring. Biologic materials should be tested for KRV infection by the MAP test. Cesarean section should be

successful for rederiving valuable breeding stocks of rats because transplacental transmission is not considered important.

Interference with Research. KRV can contaminate transplantable tumors and rat cell cultures, interfere with in vitro lymphocyte responses, suppress the development of Moloney virus-induced leukemia, and alter in vitro lymphocyte responses and cytotoxic lymphocyte activity. KRV also has been reported to induce interferon production. Immunosuppression can cause clinical disease in inapparently infected rats.

Suggested Reading

Coleman, G. L., R. O. Jacoby, P. N. Bhatt, A. L. Smith, and A. M. Jonas. 1983. Naturally occurring lethal parvovirus infection in juvenile and young adult rats. Vet. Pathol. 20:44-56.

Jacoby, R. O., P. N. Bhatt, and A. M. Jonas. 1979. Viral diseases. Pp. 273-283 in The Laboratory Rat. Vol. I: Biology and Diseases, H. J. Baker, J. R. Lindsey, and S. H. Weisbroth, eds. New York: Academic Press.

Jacoby, R. O., P. N. Bhatt, D. J. Gaertner, A. L. Smith, and E. A. Johnson. 1987. The pathogenesis of rat virus infection in infant and juvenile rats after oronasal inoculation. Arch. Virol. 95:251-270.

Jacoby, R. O., D. J. Gaertner, P. N. Bhatt, F. X. Paturzo, and A. L. Smith. 1988. Transmission of experimentally-induced rat virus infection. Lab. Anim. Sci. 38:11-14.

Kilham, L., and G. Margolis. 1966. Spontaneous hepatitis and cerebellar hypoplasia in suckling rats due to congenital infection with rat virus. Am. J. Pathol. 49:457-475.

Paturzo, F. X., R. O. Jacoby, P. N. Bhatt, A. L. Smith, D. J. Gaertner, and R. B. Ardito. 1987. Persistence of rat virus in seropositive rats as detected by explant culture. Arch. Virol. 95:137-142.

Lactic Dehydrogenase-Elevating Virus

Agent. RNA virus, family Togaviridae.

Animals Affected. Mice.

Epizootiology. Wild mice presumably serve as reservoir hosts. Transmission occurs primarily during passage of contaminated tumors, cells, or other biologic materials. Lactic dehydrogenase-elevating virus (LDV) is shed in feces, urine, saliva, and milk. After the first week of infection, the virus titer in these excretions declines sufficiently to make the risk of transmission to other mice relatively low. Transplacental transmission can occur. Bite wounds increase transmission between cagemates.

Clinical. Natural infections are usually subclinical. There is lifelong viremia in which the virus is complexed to antiviral antibody and lifelong elevation in plasma lactic dehydrogenase (LDH) and other plasma enzymes. The virus causes overt disease in C58 and AKR mice that are naturally immunosuppressed because of a loss of Lyt-1,2 cells between 5 months and 1 year of age. These mice develop polioencephalomyelitis with flaccid hind limb paralysis. C58 are more susceptible than AKR mice. Inheritance of this susceptibility is thought to be polygenic, possibly involving the *H-2* complex.

Pathology. Virus titer in serum peaks 12-14 hours postinfection and gradually decreases until about 2 weeks postinfection, at which time the titer stabilizes for life. Serum LDH increases to 8- to 11-fold above normal by 72-96 hours postinfection because of the decreased clearance of LDH V, one of the five LDH isozymes. It gradually declines over the next 3 months, although it remains significantly elevated for life. SJL/J mice carry a recessive trait that causes a 15- to 20-fold increase in serum LDH. The activity of several other serum enzymes is also increased but to a lesser degree than LDH. LDV selectively replicates in a small subpopulation of macrophages, the specific identity of which is not known. As a result, cellular immunity is depressed during the first few weeks of infection and gradually returns to normal after weeks or months. There is an enhanced humoral response to antigenic challenge with T-cell-dependent antigens during the first 24 hours postinfection but a diminished humoral response to such challenge 3 weeks or longer after infection. There is a similar enhanced response to early postinfection challenge with T-cell-independent antigen but no diminution of response in chronic infection, which suggests a defect in T-cell function. Circulating antigen-antibody complexes, which partially neutralize LDV, are produced by 4 weeks postinfection. These complexes are deposited in glomeruli, but they produce only a mild membranous glomerulopathy. Protective antibody is not produced. After 5 months of age, infected C58 and AKR mice develop age-dependent encephalomyelitis characterized by neuronal destruction, mononuclear infiltration, and microglial proliferation in the gray matter of the central nervous system.

Diagnosis. Diagnosis is usually based on the presence of increased plasma LDH activity. Transplantable tumors, virus inocula, and other biologic materials can be screened for LDV contamination by inoculating pathogen-free mice with the test material and performing a plasma or serum LDH assay 72-96 hours later. Virus isolation is not practical for most diagnostic purposes.

Control. Mice from commercial barrier breeding facilities are not likely to be infected. LDV-free mice can be derived from contaminated stocks by selection of animals with normal plasma LDH concentration or by cesarean derivation. LDV can be eliminated from tumors by passage of tumor cells in a rodent species other than the mouse or by maintenance of tumor cells in tissue culture.

Interference with Research. The effects of LDV on research results can be subtle and complex. Subclinical infection with LDV lasts throughout life, and the effects on many biologic endpoints can differ dramatically with time after infection. LDV-induced immunosuppression can cause an alteration in defense mechanisms against other infectious agents. It has been reported that tumor growth is enhanced early after LDV infection because of depressed cellular immunity and is influenced less during chronic infection. LDV infection also has been shown to alter the incidence and behavior of spontaneous virus-induced neoplasms, including the

Bittner mammary tumor and murine sarcoma viruses; to suppress the development of urethan-induced pulmonary adenomas; and to suppress vinyl chloride-vinyl acetate-induced carcinogenesis. It has been reported to cause delayed allograft rejection, to prevent the development of experimental allergic encephalomyelitis, and to prevent autoimmune disease in NZB and (NZB×NZW)F_1 hybrid mice. Serum gamma globulin levels and humoral antibody responses have been shown to increase during early infection, and humoral antibody responses have been shown to decrease during chronic infection. LDV has been found to be a polyclonal lymphocyte activator during the early stages of infection.

Suggested Reading

Brinton, M. A. 1982. Lactate dehydrogenase-elevating virus. Pp. 193-208 in The Mouse in Biomedical Research. Vol. II: Diseases, H. L. Foster, J. D. Small, and J. G. Fox, eds. New York: Academic Press.

Buxton, I. K., S. P. K. Chan, and P. G. W. Plagemann. 1988. The Ia antigen is not the major receptor for lactate dehydrogenase-elevating virus on macrophages from CBA and BALB/c mice. Virus Res. 9:205-219.

Contag, C. H., S. P. K. Chan, S. W. Wietgrefe, and P. G. W. Plagemann. 1986. Correlation between presence of lactate dehydrogenase-elevating virus RNA and antigens in motor neurons and paralysis in infected C58 mice. Virus Res. 6:195-209.

Crispens, C. G. 1964. On the epizootiology of the lactic dehydrogenase agent. J. Natl. Cancer Inst. 35:975-979.

Dillberger, J. E., P. Monroy, and N. H. Altman. 1987. Delayed increase in plasma lactic dehydrogenase activity in mouse hepatitis virus-infected mice subsequently infected with lactic dehydrogenase virus. Lab. Anim. Sci. 37:792-794.

Rowson, K. E. K., and B. W. J. Mahy. 1985. Lactate dehydrogenase-elevating virus. J. Gen. Virol. 66:2297-2312.

Leukemia Viruses, Murine

Agents. RNA virus, family Retroviridae, type C oncovirus group.

Animals Affected. Laboratory and wild mice.

Epizootiology. Murine leukemia viruses (MuLVs) are integrated into the DNA of the host's sex cells and are transmitted vertically as Mendelian traits. All laboratory and wild *Mus musculus* are thought to harbor these viruses. Horizontal transmission is inefficient but can occur through infected saliva, sputum, urine, feces, or milk or by intrauterine infection.

Clinical. Despite the fact that all mice have endogenous MuLVs, leukemias and related malignancies occur naturally in only 1-2% of most strains. Strain susceptibility is influenced by the number of MuLV gene copies in the genome and by other genes, including *Fv-1*, *Fv-2*, *In*, *nu*, *hr*, and *Ir*. The AKR strain, for example, has a high incidence of MuLV-related spontaneous thymic lymphoma, which reaches 90% by the time the mice are 9 months of age. The most common signs of thymic lymphoma are dyspnea, peripheral lymphadenopathy, and abdominal enlargement.

Pathology. Mouse strains with a high incidence of leukemia (e.g., AKR, C58, C3H/Fa) spontaneously express high titers of ecotropic MuLV in all organs throughout life. Mouse strains with a low incidence of leukemia (e.g., BALB/c, A/J, C3H/He, CBA/J) express only low titers of virus. The mouse leukemias are, with few exceptions, actually lymphomas because they are predominantly solid tumors of lymphocytes or other hematopoietic cells. The majority of those occurring before 1 year of age are thymic lymphomas, while those seen in older mice are predominantly histiocytic (reticulum cell) lymphomas. Other types include nonthymic lymphomas, lymphatic leukemias, granulocytic leukemias, erythroleukemia, plasma cell tumors, and mast cell tumors.

Diagnosis. Diagnosis of neoplasia is based on morphologic features in histologic sections. Tumor cell types can also be determined by analysis of cell-surface antigens. Isolation and characterization of MuLVs require specialized techniques usually available only in research laboratories dedicated to viral oncology. MuLV type and group specificity can be determined by immunofluorescence of fixed cells using an appropriate panel of type- or group-specific antisera.

Control. Control measures are not usually considered useful since all mice probably have endogenous MuLV proviruses as an integral part of their genomes. Horizontal transmission is considered inefficient.

Interference with Research. The presence of MuLVs probably has little significance for most research purposes. MuLV expression and the associated occurrence of neoplasms, however, can present competing endpoints in some studies, e.g., studies of aging processes in various organs. Thus, an awareness of the incidence of spontaneous tumors in different mouse strains or the ability of specific test chemicals to induce tumors can be useful in designing some experiments. Active MuLV infection can cause suppression of humoral and cellular immunity without causing clinical disease.

Suggested Reading

Bishop, J. M., and H. E. Varmus. 1975. The molecular biology of RNA viruses. Pp. 1-48 in Cancer: A Comprehensive Treatise, vol. 2., F. F. Becker, ed. New York: Plenum.

Dunn, T. B. 1954. Normal and pathologic anatomy of reticular tissue in laboratory mice with classification and discussion of neoplasms. J. Natl. Cancer Inst. 14:1281-1433.

Furmanski, P., and M. A. Rich. 1982. Neoplasms of the hematopoietic system. Pp. 352-371 in The Mouse in Biomedical Research. Vol. IV: Experimental Biology and Oncology, H. L. Foster, J. D. Small, and J. G. Fox, eds. New York: Academic Press.

Lilly, F., and A. Mayer. 1980. Genetic aspects of murine type-C viruses and their hosts in oncogenesis. Pp. 89-108 in Viral Oncology, G. Klein, ed. New York: Raven Press.

Risser, R., J. M. Horowitz, and J. McCubrey. 1983. Endogenous mouse leukemia viruses. Annu. Rev. Genet. 17:85-121.

Squire, R. A., D. G. Goodman, M. G. Valerio, T. Frederickson, J. D. Strandberg, M. H. Levitt, C. H. Lingeman, J. C. Harshbarger, and C. J. Dawe. 1978. Pp. 1051-1283 in Pathology of Laboratory Animals, vol. II, K. Benirschke, F. M. Garner, and T. C. Jones, eds. New York: Springer-Verlag.

Lymphocytic Choriomeningitis Virus

Agent. RNA virus, family Arenaviridae, genus *Arenavirus*.

Animals Affected. Wild mice are the principal reservoir hosts, but laboratory mice and Syrian hamsters also serve as important natural hosts. Humans, monkeys, dogs, rabbits, guinea pigs, rats, and chickens are susceptible to the virus.

Epizootiology. Natural lymphocytic choriomeningitis virus (LCMV) infections of laboratory mice are rare. Only infected mice and hamsters are known to transmit the virus. Both species can have chronic infections, with high concentrations of virus shed in the urine, saliva, and milk. The portals of entry are probably mucus membranes and broken skin. Vertical (transovarian and/or transuterine) transmission occurs in mice and probably in hamsters. Once introduced into a population of mice, the infection can spread to all members of that population.

Since 1960 three epidemics involving at least 236 human cases have occurred in the United States, and all have been associated with Syrian hamsters, either as laboratory animals bearing transplantable tumors or as pets.*

Clinical. Clinical signs are highly variable depending on the virus strain, the mouse strain, and the age of the mouse at the time of infection. Persistent tolerant infection is acquired if infection occurs in utero or within a few days of birth. There is lifelong viremia and shedding of virus. Transient runting can occur during the first 3 weeks of life. Thereafter, the mice appear normal. At 7-10 months of age, immune complex glomerulonephritis occurs and is associated with emaciation, ruffled fur, hunched posture, ascites, and sometimes death. Nontolerant (acute) infection is acquired if infection occurs after the first week of life when the animals are immunocompetent. Viremia occurs, but there is no shedding of virus. The outcome is either death within a few days or weeks or recovery with elimination of the virus. Natural infections in adult mice range from inapparent infection to severe disease with high mortality. Natural infections in hamsters are usually subclinical.

Pathology. T-cell, but not B-cell, activity is suppressed in persistent tolerant infection. Infectious virus circulates bound to LCMV-specific IgG and complement. These complexes accumulate in the renal glomeruli; choroid plexus; and, to a lesser degree, in synovial membranes, in blood vessel walls, and beneath the epidermis of the skin to cause late-onset disease that becomes clinically apparent around 7-10 months of age. There is generalized lymphoid hyperplasia and perivascular accumulation of lymphocytes and plasma cells in all visceral organs. In nontolerant (acute) infection, there is multifocal hepatic necrosis and generalized necrosis of lymphoid tissues. Both the morphologic lesions and elimination of the virus are due to cell-mediated immune responses involving *H-2*-restricted, cytotoxic T lymphocytes and possibly natural killer cells. LCMV-infected adult mice can be protected from disease by numerous immunosuppressive regimens. Athymic (*nu/*

---

*In humans the usual clinical manifestations are those of flu-like disease, with fever, headache, myalgia, nausea, vomiting, sore throat, and photophobia being the major symptoms. The following occur rarely: rash, alopecia, diarrhea, cough, arthritis, lymphadenopathy, orchitis, delirium, amnesia, meningitis.

nu) mice inoculated with the virus at 3-6 weeks of age do not develop disease but become persistently viremic.

Diagnosis. The serologic methods of choice are the IFA test, the micro plaque-reduction test for neutralizing antibody, and the ELISA. The IFA test is particularly useful for rapid diagnosis early in the course of infection, while the micro plaque-reduction test is considered better for chronic infection. The CF test is considered relatively insensitive and is not recommended. Drawing and processing of blood from an animal suspected of LCMV infection should be done with care because of the likelihood of viremia.

The MAP test can be used in testing transplantable tumors and other biologic materials. The virus can also be identified by using the IFA test in tissues or cultures of isolated virus.

Control. The most practical method of control is to obtain mice only from colonies known to be free of LCMV and to maintain them in a barrier facility that excludes wild rodents. Biologic materials such as transplantable tumors should be pretested and shown to be free of the virus before experimental use. If LCMV infection is diagnosed, the entire stock should be destroyed and incinerated. Animal cages and other equipment should be autoclaved. Animal rooms should be fumigated either with formalin (40% formaldehyde in water, sprayed on all room surfaces at 36 ml/m^3) or paraformaldehyde (11 g/m^3 vaporized in a high-temperature silicone fluid at 96°C) and allowed to remain vacant for 7-10 days. Cesarean derivation of animal stocks is of no value because of transovarian or transuterine transmission of infection.

Interference with Research. LCMV infection is an important zoonotic infection that can cause serious disease and sometimes fatality in personnel. LCMV is a frequent contaminant of biologic materials, including transplantable tumors of mice, hamsters, and guinea pigs; tissue culture cell lines; virus stocks, including leukemia viruses, distemper virus, rabies virus, and mouse poliomyelitis virus; and *Toxoplasma gondii* sublines. LCMV infection has an inhibitory effect on tumor induction by polyomavirus, Rauscher virus, and mammary tumor virus. Acute infection has been reported to cause induction of natural killer cell activity, depression of both humoral and cellular immunity, delayed rejection of skin and tumor allografts, increased susceptibility to mousepox virus or *Eperythrozoon coccoides* infection, and increased susceptibility to bacterial endotoxin and x-irradiation. Abrogation of the naturally occurring insulin-dependent diabetes mellitus of BB strain rats has also been reported. Chronic infection has been shown to cause proliferation of virus-specific cytotoxic T lymphocytes.

Suggested Reading

Bhatt, P. N., R. O. Jacoby, and S. W. Barthold. 1986. Contamination of transplantable murine tumors with lymphocytic choriomeningitis virus. Lab. Anim. Sci. 36:136-139.

Biggar, R. J., R. Deibel, and J. P. Woodall. 1976. Implications, monitoring, and control of accidental transmission of lymphocytic choriomeningitis virus within hamster tumor cell lines. Cancer Res. 36:537-553.

Lehmann-Grube, F. 1982. Lymphocytic choriomeningitis virus. Pp. 231-266 in The Mouse in Biomedical Research. Vol. II: Diseases, H. L. Foster, J. D. Small, and J. G. Fox, eds. New York: Academic Press.

Lohler, J., and F. Lehmann-Grube. 1981. Immunopathologic alterations of lymphatic tissues of mice infected with lymphocytic choriomeningitis virus. I. Histopathologic findings. Lab. Invest. 44:193-204.

Moskophidis, D., S. P. Coppold, H. Waldmann, and F. Lehmann-Grube. 1987. Mechanism of recovery from acute virus infection: Treatment of lymphocytic choriomeningitis virus-infected mice with monoclonal antibodies reveals that Lyt-2^+ T lymphocytes mediate clearance of virus and regulate the antiviral antibody response. J. Virol. 61:1867-1874.

van der Zeijst, B. A. M., B. E. Noyes, M.-E. Mirault, B. Parker, A. D. M. E. Osterhaus, E. A. Swyryd, N. Bleumink, M. C. Horzinek, and G. R. Stark. 1983. Persistent infection of some standard cell lines by lymphocytic choriomeningitis virus: Transmission of infection by an intracellular agent. J. Virol. 48:249-261.

Mammary Tumor Virus, Mouse

Agent. RNA virus, family Retroviridae. Four major variants have been identified: MMTV-S (standard; the Bittner virus), which is transmitted through the milk to nursing young and is highly oncogenic; MMTV-L (low oncogenic), which is transmitted through germ cells and is weakly oncogenic; MMTV-P (pregnancy-dependent), which is transmitted through both milk and germ cells and is highly oncogenic; MMTV-O (overlooked), which is considered an endogenous virus in the genome of most mice.

Animals Affected. Wild and laboratory mice.

Epizootiology. When infected, strains such as C3H, DBA/2, and A readily express MMTV-S, and the virus can be demonstrated in a variety of locations throughout the body, especially in mammary tissue and milk. Transmission of MMTV-S is by ingestion of infected milk, resulting in a high incidence of mammary tumors early in life (6-12 months) when the associated genetic and hormonal factors are also present.

Clinical. Tumors occur in any part of the body where mammary tissue is located. The lungs are the most common site for distant metastases.

Pathology. It is thought that the virus initially induces hyperplastic alveolar nodules that progress to neoplasia. The average latency period from infection to tumor expression is 6-9 months. Susceptibility to MMTV is genetically determined, and tumor development is enhanced by administration of estrogen to both males and females, forced breeding, and administration of carcinogens. Mammary tumors are usually circumscribed, round to nodular, gray to white masses located in the subcutaneous tissue. Ulcerations and hemorrhages are common in large tumors. Histologically, most are types A or B adenocarcinomas. Type A tumors are characterized by uniform acini lined by a single layer of cuboidal cells. Type B tumors are variable in the extent of differentiation but usually consist of irregular cords and sheets of cells. Types C, Y, L, and P adenocarcinomas; carcinomas with squamous cell differentiation; and carcinosarcomas occur less frequently. Other histologic types are rare. Mice of many strains develop humoral and cellular immune responses to MMTV, indicating that mice infected early in life are not immunologically tolerant.

Diagnosis. Pathologic diagnosis of mouse mammary tumors is based on histopathologic characteristics. Detection and characterization of the virus require test procedures normally available only in specialized viral oncology laboratories, including nucleic acid hybridization, immunologic assays for viral antigens, and bioassays for infectivity in different strains of mice.

Control. The most practical method of control is by selection of mouse strains without MMTV. Foster nursing of young on mouse strains that are free of the virus has been used to eliminate MMTV-S.

Interference with Research. MMTV infection can be a complicating factor in experimental carcinogenesis studies.

Suggested Reading

Bentvelzen, P., and J. Hilgers. 1980. Murine mammary tumor virus. Pp. 311-355 in Viral Oncology, G. Klein, ed. New York: Raven Press.

Bittner, J. J. 1936. Some possible effects of nursing on the mammary gland tumor incidence in mice. Science 84:162.

Medina, D. 1982. Mammary tumors. Pp. 373-396 in The Mouse in Biomedical Research. Vol. IV: Experimental Biology and Oncology, H. L. Foster, J. D. Small, and J. G. Fox, eds. New York: Academic Press.

Michalides, R., A. van Ooyen, and R. Nusse. 1983. Mammary tumor virus expression and mammary tumor development. Curr. Top. Microbiol. Immunol. 106:57-78.

Ringhold, G. M. 1983. Regulation of mouse mammary tumor virus gene expression by glucocorticoid hormones. Curr. Top. Microbiol. Immunol. 106:79-103.

Sass, B., and T. B. Dunn. 1979. Classification of mouse mammary tumors in Dunn's miscellaneous group including recently reported types. J. Natl. Cancer Inst. 62:1287-1293.

Minute Virus of Mice

Agent. DNA virus, family Parvoviridae, genus *Parvovirus*.

Animals Affected. Wild and laboratory mice.

Epizootiology. Wild mice serve as reservoir hosts. Enzootic infection is common in barrier-maintained and conventional breeding colonies of mice. Minute virus of mice (MVM) is highly contagious. In infected colonies, maternal antibodies are protective until the young are 6-8 weeks of age; however, most mice become infected and seroconvert by 3 months of age. Transmission occurs by direct contact and by urine and fecal contamination. Airborne and transplacental infection are not considered important.

Clinical. Natural infections are inapparent.

Pathology. There are no pathologic changes associated with natural infections.

Diagnosis. The ELISA and the IFA test are the most sensitive serologic tests. The HAI, CF, and neutralization tests are used to discriminate infections caused by MVM from those caused by Kilham rat virus or H-1 virus. MRL/MpJ and MRL/MpJ-*lpr/lpr* mice frequently show false-positive HAI test results for MVM. Virus isolation can be done by using rat embryo tissue culture. The MAP test is commonly used for detection of MVM in transplantable tumors and other biologic materials.

Control. Infection can be eliminated from stocks of mice by cesarean derivation, but elimination of infected mice followed by replacement with MVM-free mice is often more practical. Strict adherence to barrier procedures is required to maintain the MVM-free state. Wild mice must be excluded. Transplantable tumors, virus stocks, and other biologic materials should be monitored before admission to a facility.

Interference with Research. MVM is a frequent contaminant of mouse leukemia virus preparations, transplantable tumors, hybridomas, and cell lines. Evidence that MVM can interfere with research has come from studies of MVM(i), a single variant of the virus that may or may not occur as a natural infection in contemporary mice. MVM(i) grows lytically in cytotoxic T-lymphocyte clones, abrogates cytotoxic T-lymphocyte responses, suppresses T-lymphocyte mitogenic responses, and suppresses T-helper-dependent B-lymphocyte responses in vitro. The intramuscular inoculation of MVM(p) into mice suppresses the growth of Ehrlich ascites tumor cells given intraperitoneally.

Suggested Reading

Cross, S. S., and J. C. Parker. 1972. Some antigenic relationships of the murine parvoviruses: Minute virus of mice, rat virus, and H-1 virus. Proc. Soc. Exp. Biol. Med. 139:105-108.

Engers, H. D., J. A. Louis, R. H. Zubler, and B. Hirt. 1981. Inhibition of T cell mediated functions by MVM(i), a parvovirus closely related to minute virus of mice. J. Immunol. 127:2280-2285.

Parker, J. C., S. S. Cross, M. J. Collins, Jr., and W. P. Rowe. 1970. Minute virus of mice. I. Procedures for quantitation and detection. J. Natl. Cancer Inst. 45:297-303.

Parker, J. C., M. J. Collins, Jr., S. S. Cross, and W. P. Rowe. 1970. Minute virus of mice. II. Prevalence, epidemiology, and occurrence as a contaminant of transplanted tumors. J. Natl. Cancer Inst. 45:305-310.

Tattersall, P., and S. F. Cotmore. 1986. The rodent parvoviruses. Pp. 305-348 in Viral and Mycoplasmal Infections of Laboratory Rodents: Effects on Biomedical Research, P. N. Bhatt, R. O. Jacoby, H. C. Morse III., and A. E. New, eds. Orlando, Fla.: Academic Press.

Ward, D. C., and P. J. Tattersall. 1982. Minute virus of mice. Pp. 313-334 in The Mouse in Biomedical Research. Vol. II: Diseases, H. L. Foster, J. D. Small, and J. G. Fox, eds. New York: Academic Press.

Mycoplasma arthritidis

Agent. Gram-negative bacterium, family Mycoplasmataceae.

Animals Affected. Rats and mice.

Epizootiology. *M. arthritidis* occurs as a common subclinical infection in rats and mice, including cesarean-derived, barrier-maintained stocks.

Clinical. The infection is usually subclinical; disease caused by this agent is extremely rare.

Pathology. There are no pathologic changes associated with subclinical infections.

Diagnosis. The ELISA should be used for screening; immunoblot on positive sera can be used to discriminate between species of *Mycoplasma*. Isolation of *M. arthritidis* from animals with subclinical infection requires culture of tissue homogenates from multiple organ sites, which is not practical in most instances. In the rare event that clinical arthritis occurs, the organism should be cultured from joint exudates.

Control. Definitive information is not available.

Interference with Research. Infection can cause spontaneous polyarthritis in rats. Subclinical infections can be activated to complicate experimentally induced arthritis in rats. *M. arthritidis* can contaminate transplantable tumors of rats, causing arthritis and/or abscesses at the injection site in recipients. Experimental infections of rodents with *M. arthritidis* can be complicated by preexisting latent infection with this organism. *M. arthritidis* is a frequent contaminant of rodent cell cultures.

Suggested Reading

Cassell, G. H., M. K. Davidson, J. K. Davis, and J. R. Lindsey. 1983. Recovery and identification of murine mycoplasmas. Pp. 129-142 in Methods in Mycoplasmology, vol. II, J. G. Tully and S. Razin, eds. New York: Academic Press.

Davidson, M. K., J. R. Lindsey, M. B. Brown, G. H. Cassell, and G. H. Boorman. 1983. Natural infection of *Mycoplasma arthritidis* in mice. Curr. Microbiol. 8:205-208.

Lindsey, J. R., M. K. Davidson, T. R. Schoeb, and G. H. Cassell. 1986. Murine mycoplasmal infections. Pp. 91-121 in Complications of Viral and Mycoplasma Infections in Rodents to Toxicology Research and Testing, T. E. Hamm, Jr., ed. Washington, D.C.: Hemisphere.

Minion, F. C., M. B. Brown, and G. H. Cassell. 1984. Identification of cross-reactive antigens between *Mycoplasma pulmonis* and *Mycoplasma arthritidis*. Infect. Immun. 43:115-121.

Razin, S., and E. A. Freundt. 1984. The mycoplasmas. Pp. 740-793 in Bergey's Manual of Systematic Bacteriology, vol. I, N. R. Krieg and J. G. Hold, eds. Baltimore: Williams & Wilkins.

Thirkill, C. E., and D. S. Gregerson. 1982. *Mycoplasma arthritidis*-induced ocular inflammatory disease. Infect. Immun. 36:775-781.

Mycoplasma pulmonis

Agent. Gram-negative bacterium, family Mycoplasmataceae.

Animals Affected. Rats and mice. The organism is occasionally isolated from wild rats (*Rattus norvegicus*), cotton rats (*Sigmodon hispidus hispidus*), rabbits, Syrian hamsters, and guinea pigs.

Epizootiology. Infection and disease are common in conventionally reared rats and mice. Subclinical infection occurs in some cesarean-derived, barrier-maintained stocks. Transmission is thought to be by the intrauterine route and by aerosol between cagemates, including from dam to offspring and between adjacent cages.

Clinical. Infections are usually subclinical. Signs of murine respiratory mycoplasmosis (MRM), the disease caused by *M. pulmonis*, are nonspecific but can include rales, polypnea, weight loss, hunched posture, ruffled coat, inactivity, and "head tilt," in both rats and mice; "snuffling" and accumulation of porphyrin pigment around the eyes and external nares in rats; and "chattering" in mice. Athymic (*nu/nu*) mice are no more susceptible to MRM than are immunocompetent mice.

Pathology. *M. pulmonis* is an extracellular parasite that preferentially colonizes the luminal surface of respiratory epithelium. Organisms and lesions, if present, tend to decrease from proximal to distal airways. Usually, the organism is a commensal. Intracage ammonia concentrations of 19 µg/l of air or greater appear to exacerbate MRM by increasing the growth of *M. pulmonis* in the respiratory tract. Other influencing factors include concurrent infection with Sendai virus, sialodacryoadenitis virus or cilia-associated respiratory bacillus; administration of hexamethylphosphoramide or cyclophosphamide; a deficiency of vitamins A or E; inhalation of tobacco smoke; genetic susceptibility of the host (e.g., LEW are more susceptible than F344 rats; C3H/HeN are more susceptible than C57BL/6N mice); and, possibly, virulence of the *M. pulmonis* strain. Characteristic changes at any level in the respiratory tract include neutrophils in the airways, hyperplasia of the mucosal epithelium, and a lymphoid response in the submucosa. Lesions can be acute or chronic and include rhinitis, otitis media, laryngitis, tracheitis, bronchitis, bronchiectasis, pulmonary abscesses, and alveolitis. Pleuritis and emphysema are rare. Hyperplasia of bronchus-associated lymphoid tissue is characteristic in rats. Syncytial epithelial giant cells can occur in nasal and bronchial mucosa in mice. LEW rats show genital disease, characterized by purulent endometritis, pyometra, salpingitis, and perioophoritis. In mice humoral antibody is protective and can be passively transferred. Cellular immunity appears to be more important in rats than in mice.

Diagnosis. Cultural isolation can be achieved by using a medium that has been pretested and shown to support growth. The nasopharynx is probably the best single site from which to obtain a culture sample, but culturing samples from multiple sites increases the isolation rate. A battery of bacterial, viral, and histopathologic procedures should be used to identify the responsible agent(s) and to exclude other possible causes or contributors. Efforts should be made to identify factors that exacerbate MRM. The ELISA is the method of choice for rodent health surveillance; it is more sensitive and cost effective than is culturing the organism. The ELISAs currently in use are only genus specific; therefore, immunoblot is used to differentiate between species of *Mycoplasma*. Detection of subclinical infection is often a major problem; ELISA seropositivity might occur only sporadically, the number of ELISA-positive animals might be very small, and seropositive animals might become negative again. The best results are obtained when only adults are tested (weanlings with subclinical infection usually are ELISA negative), sample size is increased, and testing is done repeatedly.

Control. Cesarean-derivation and barrier-maintenance programs appear to have reduced the prevalence of disease but may not have been equally successful in reducing the prevalence of infection. The major emphasis should be on selecting mycoplasma-free breeding stocks. This can be achieved by housing small groups of young adult breeders in plastic film isolators and testing them monthly using the ELISA until they are 12 months of age. Young animals from stocks found to be consistently negative can then be used to establish breeder production populations under barrier programs. Definitive information on eliminating *M. pulmonis* from clinically or subclinically infected stocks is lacking. Dams should be several months old and have been found repeatedly to be ELISA negative. Administration of antimicrobial agents might help to control clinical signs; however, such agents are not curative and can introduce variables if used in animals on experimental protocols.

Interference with Research. Morbidity and mortality caused by MRM can disrupt long-term studies. MRM alters ciliary function, cell kinetics, and immunity in the respiratory tract and changes the response to carcinogens. Infection in LEW rats delays the onset and reduces the severity of adjuvant arthritis, reduces the incidence of experimental collagen-induced arthritis, and reduces antibody response to collagen. Genital infection alters genital tract histology. In mice infection can activate natural killer cells, contaminate transplantable tumors, and cause arthritis in the recipients. Subclinical infection can be exacerbated by some experimental procedures (e.g., deficiencies of vitamin A or E, administration of hexamethylphosphoramide). *M. pulmonis* is a frequent contaminant of rodent cell cultures. Mycoplasmas produce lymphokine-like substances that are mitogenic for B and T lymphocytes in vitro.

Suggested Reading

Cassell, G. H., N. R. Cox, J. K. Davis, M. B. Brown, F. C. Minion, and J. R. Lindsey. 1986. State-of-the-art detection methods for rodent mycoplasmas. Pp. 143-160 in Complications of Viral and Mycoplasmal Infections in Rodents to Toxicology Research and Testing, T. E. Hamm, Jr., ed. Washington, D.C.: Hemisphere.

Cassell, G. H., M. K. Davidson, J. K. Davis, and J. R. Lindsey. 1983. Recovery and identification of murine mycoplasmas. Pp. 129-142 in Methods in Mycoplasmology, vol. II, J. G. Tully and S. Razin, eds. New York: Academic Press.

Lindsey, J. R., G. H. Cassell, and H. J. Baker. 1978. Mycoplasmatales and rickettsiales. Pp. 1481-1550 in Pathology of Laboratory Animals, vol. II, K. Benirschke, F. Garner, and C. Jones, eds. New York: Springer-Verlag.

Lindsey, J. R., M. K. Davidson, T. R. Schoeb, and G. H. Cassell. 1986. Murine mycoplasmal infections. Pp. 91-121 in Complications of Viral and Mycoplasma Infections in Rodents to Toxicology Research and Testing, T. E. Hamm, Jr., ed. Washington, D.C.: Hemisphere.

Minion, F. C., M. B. Brown, and G. H. Cassell. 1984. Identification of cross-reactive antigens between *Mycoplasma pulmonis* and *Mycoplasma arthritidis*. Infect. Immun. 43:115-121.

Razin, S., and E. A. Freundt. 1984. The mycoplasmas. Pp. 740-793 in Bergey's Manual of Systematic Bacteriology, vol. I, N. R. Krieg and J. G. Holt, eds. Baltimore: Williams & Wilkins.

Pasteurella pneumotropica

Agent. Gram-negative coccobacillus, family Pasteurellaceae.

Animals Affected. Mice, rats, hamsters, guinea pigs, and many others.

Epizootiology. *P. pneumotropica* can be isolated from up to 95% of healthy animals in some colonies. It can be isolated from many organs, including the respiratory tract, oral cavity, intestine, uterus, urinary bladder, skin, and conjunctiva. Transmission is probably by contact and fomites.

Clinical. Infections are usually subclinical. When clinical infections do occur, signs in mice include conjunctivitis, panophthalmitis, dacryoadenitis, subcutaneous and cervical abscesses, bulbourethral gland infections, uterine infections, and otitis media, while signs in rats include ophthalmitis, conjunctivitis, subcutaneous abscesses, and mastitis.

Pathology. *P. pneumotropica* is an opportunist that most frequently causes lesions of the skin and adnexal structures. Lesions are usually characterized by suppurative inflammation.

Diagnosis. Diagnosis must discriminate between *P. pneumotropica* infection and *P. pneumotropica*-induced disease and rule out other possible causative agents and disease processes. *Pasteurella* spp., *Actinobacillus* spp., *Haemophilus* spp., and *Yersinia* spp., which are commonly found in mice and rats, give similar reactions in many biochemical tests. Therefore, extensive biochemical testing is required to accurately identify these organisms. An ELISA for detection of serum antibody to *P. pneumotropica* has recently been developed.

Control. Cesarean derivation and maintenance in a gnotobiotic isolator may be necessary to exclude the organism completely. Antibiotic therapy has limited value.

Interference with Research. There have been no reports of interference with research results.

Suggested Reading

Carter, G. R. 1984. Genus 1. *Pasteurella*. Pp. 552-558 in Bergey's Manual of Systematic Bacteriology, vol. I, N. R. Krieg and J. G. Holt, eds. Baltimore: Williams & Wilkins.

Jawetz, E. 1950. A pneumotropic *Pasteurella* of laboratory animals. I. Bacteriological and serologic characteristics of the organism. J. Infect. Dis. 86:172-183.

Jawetz, E., and W. H. Baker. 1950. A pneumotropic *Pasteurella* of laboratory animals. II. Pathological and immunological studies with the organism. J. Infect. Dis. 86:184-196.

Kunstyr, I., and D. Hartman. 1983. *Pasteurella pneumotropica* and the prevalence of the AHP (*Actinobacillus, Haemophilus, Pasteurella*)-group in laboratory animals. Lab. Anim. (London) 17:156-160.

Moore, T. D., A. M. Allen, and J. R. Ganaway. 1973. Latent *Pasteurella pneumotropica* infection of the intestine of gnotobiotic and barrier-held rats. Lab. Anim. Sci. 23:657-661.

Pneumocystis carinii

Agent. Classification uncertain. Considered a fungus or a protozoan.

Animals Affected. Mice, rats, humans, and numerous other mammals.

Epizootiology. *P. carinii* is a ubiquitous organism of low virulence. It causes active pulmonary disease only in immunocompromised hosts. It is extremely prevalent as a persistent subclinical infection in mice and rats. Transmission is thought to be by inhalation of infective cysts expelled during exhalation or coughing.

Clinical Signs. Rats and mice show no clinical signs unless they are immunodeficient or immunosuppressed. Clinical signs include weight loss, cyanosis, rough hair coat, and dyspnea.

Pathology. In animals with active disease the lungs are enlarged, rubbery in consistency, plum colored, and heavier than normal. Histologically, alveolar septae are variably thickened, and there is a meager inflammatory response of lymphoid cells. Many alveoli are distended by homogeneous, foamy, eosinophilic material characteristic of *P. carinii* pneumonia.

Diagnosis. Definitive identification of *P. carinii* in active infection depends on the demonstration of cysts (characteristically containing eight sporozoites) and trophozoites. Cysts measure 5-7 µm in diameter and have thick walls that stain with methenamine silver, cresyl violet, periodic acid-Schiff, or toluidine blue. Giemsa stain is preferred for the demonstration of trophozoites and sporozoites in lung imprints. An IFA method also has been used. Acridine orange has been proposed for rapid screening of imprints; trophozoites stain yellow to orange; cyst walls do not stain. Persistent infections can be diagnosed by immunosuppressing some animals and then demonstrating the organisms in diseased lung tissue.

Control. Subclinical infection is probably very common in conventionally reared and pathogen-free colonies. Gnotobiotic methods are likely useful in excluding the infection but might not be completely effective because of possible vertical transmission.

Interference with Research. *P. carinii* can complicate long-term studies in which severely immunosuppressed animals are used. Athymic (*nu/nu*) and severe combined immunodeficient (*scid/scid*) mice can develop active infection.

Suggested Reading

Bartlett, M. S., M. M. Durkin, M. A. Jay, S. F. Queener, and J. W. Smith. 1987. Sources of rats free of latent *Pneumocystis carinii*. J. Clin. Microbiol. 25:1794-1795.

Walzer, P. D., R. D. Powell, and K. Yoneda. 1979. Experimental *Pneumocystis carinii* pneumonia in different strains of cortisonized mice. Infect. Immun. 24:939-947.

Walzer, P. D., R. Powell, Jr., K. Yoneda, M. E. Rutledge, and J. E. Milder. 1980. Growth characteristics and pathogenesis of experimental *Pnemocystis carinii* pneumonia. Infect. Immun. 27:928-937.

Walzer, P. D., M. E. Rutledge, and K. Yoneda. 1983. Experimental *Pneumocystis carinii* pneumonia in C3H/HeJ and C3HeB/FeJ mice. J. Reticuloendothel. Soc. 33:1-9.

Walzer, P. D., C. K. Kim, J. Linke, C. L. Pogue, M. J. Huerkamp, C. E. Chrisp, A. V. Lerro, S. K. Wixson, E. Hail, and L. D. Shultz. 1989. Outbreaks of *Pneumocystis carinii* pneumonia in colonies of immunodeficient mice. Infect. Immun. 57:62-70.

Weir, E. C., D. G. Brownstein, and S. W. Barthold. 1986. Spontaneous wasting disease in nude mice associated with *Pneumocystis carinii* infection. Lab. Anim. Sci. 36:140-144.

Pneumonia Virus of Mice

Agent. RNA virus, family Paramyxoviridae, genus *Pneumovirus*.

Animals Affected. Mice, rats, and hamsters.

Epizootiology. Prevalence rates are approximately 50% in rat and hamster colonies and 20% in mouse colonies. Active infection in mice (and presumably in rats and hamsters) lasts about 9 days. Chronic or latent infections do not occur. Transmission is exclusively horizontal via the respiratory tract, mainly by direct contact and aerosols. Fomites are probably not important in transmission.

Clinical Signs. Natural infections are subclinical, except in immunocompromised hosts. Chronic illness, emaciation, and death have been reported in infected athymic (*nu/nu*) mice.

Pathology. No pathologic lesions have been associated with natural infections in immunocompetent hosts. Chronic pneumonia has been reported to occur in naturally infected athymic mice.

Diagnosis. The ELISA is the most sensitive method for routine monitoring. The HAI test is also reliable, although occasionally it gives false-positive results. The CF test is useful for detecting recent PVM infections. The MAP test is useful for detecting the virus in biologic specimens. The virus can be isolated by using primary hamster kidney, BHK-21, Vero, or hamster embryo cells.

Control. Cesarean derivation and barrier maintenance are effective methods of controlling infection.

Interference with Research. There have been no reports of interference with research results.

Suggested Reading

Carrano, V. A., S. W. Barthold, D. S. Beck, and A. L. Smith. 1984. Alteration of viral respiratory infections of mice by prior infection with mouse hepatitis virus. Lab. Anim. Sci. 34:573-576.

Jacoby, R. O., P. N. Bhatt, and A. M. Jonas. 1979. Viral diseases. Pp. 271-306 in The Laboratory Rat. Vol. I: Biology and Diseases, H. J. Baker, J. R. Lindsey, and S. H. Weisbroth, eds. New York: Academic Press.

Parker, J. C., and C. B. Richter. 1982. Viral diseases of the respiratory system. Pp. 107-152 in The Mouse in Biomedical Research. Vol. II: Diseases, H. L. Foster, J. D. Small, and J. G. Fox, eds. New York: Academic Press.

Richter, C. B., J. E. Thigpen, C. S. Richter, and J. M. MacKenzie, Jr. 1988. Fatal pneumonia with terminal emaciation in nude mice caused by pneumonia virus of mice. Lab. Anim. Sci. 38:255-261.

Smith, A. L., V. A. Carrano, and D. G. Brownstein. 1984. Response of weanling random-bred mice to infection with pneumonia virus of mice (PVM). Lab. Anim. Sci. 34:35-37.

Weir, E. C., D. G. Brownstein, A. L. Smith, and E. A. Johnson. 1988. Respiratory disease and wasting in athymic mice infected with pneumonia virus of mice. Lab. Anim. Sci. 38:133-137.

Polyomavirus

Agent. DNA virus, family Papovaviridae, genus *Polyomavirus*.

Animals Affected. Wild and laboratory mice.

Epizootiology. Polyomavirus is very rare in mice from commercial barrier breeding facilities in the United States. The virus is highly contagious and is shed in large quantities in saliva, urine, and feces of infected mice. In persistently infected dams, the titer of virus in the kidney increases during late pregnancy. The virus has a propensity for airborne dissemination and intranasal infection. Contaminated feed and bedding can also be important sources of infection.

Clinical. Natural infections are usually inapparent. Athymic (*nu/nu*) mice given heterotransplants of human tumors contaminated with the virus have been reported to develop a syndrome characterized by wasting and paralysis of the rear legs and tail.

Pathology. Morphologic lesions, including polyomavirus-induced tumors, are usually not found in naturally infected mice.

Diagnosis. The ELISA and HAI test are commonly used in routine health monitoring. The MAP test can be used for screening tumor lines and other biologic materials. Virus isolation using mouse embryo tissue culture can be useful in selected situations.

Control. Cesarean derivation and barrier maintenance are usually effective in eliminating the virus because transplacental transmission does not occur. Strict isolation from polyomavirus-infected stocks and exclusion of wild mice from animal facilities are essential for preventing infection. Laminar-flow units and filter-top cages are helpful in reducing the spread of infection in laboratories in which the agent is used experimentally.

Interference with Research. Polyomavirus can complicate research by contaminating tumor lines, stocks of other viruses, and other biologic materials that are passaged in mice. Polyomavirus-contaminated tumors can cause paralysis and wasting in athymic (*nu/nu*) mice.

Suggested Reading

Eddy, B. E. 1982. Polyomavirus. Pp. 293-311 in The Mouse in Biomedical Research. Vol. II: Diseases, H. L. Foster, J. D. Small, and J. G. Fox, eds. New York: Academic Press.

Harper, J. S., III, C. J. Dawe, B. D. Trapp, P. E. McKeever, M. Collins, J. L. Woyclechowska, D. L. Madden, and J. L. Sever. 1983. Paralysis in nude mice caused by polomavirus-induced vertebral tumors. Pp. 359-367 in Polyomaviruses and Human Neurological Diseases. New York: Alan R. Liss.

McCance, D. J., A. Sebesteny, B. E. Griffin, F. Balkwill, R. Tilly, and N. A. Gregson. 1983. A paralytic disease in nude mice associated with polyoma virus infection. J. Gen. Virol. 64:57-67.

McGarrity, G. J., and A. S. Dion. 1978. Detection of airborne polyoma virus. J. Hyg. 81:9-13.

McGarrity, G. J., L. L. Coriell, and V. Ammen. 1976. Airborne transmission of polyomavirus. J. Natl. Cancer Inst. 56:159-162.

Shah, K. V., and C. Christian. 1986. Mouse polyoma and other papovaviruses. Pp. 505-527 in Viral and Mycoplasmal Infections of Laboratory Rodents: Effects on Biomedical Research, P. N. Bhatt, R. O. Jacoby, H. C. Morse III, and A. E. New, eds. Orlando, Fla.: Academic Press.

Pseudomonas aeruginosa

Agent. Gram-negative bacterium, family Pseudomonadaceae.

Animals Affected. Mice, rats, humans, and numerous other species.

Epizootiology. *P. aeruginosa* is ubiquitous, occurring widely in soil, water, sewage, and air. It is widely distributed in conventional stocks of rodents and is transmitted by fomites (e.g., contaminated food, bedding, or water) or by contact with infected humans or rodents.

Clinical. The organism is sometimes part of the normal flora in the digestive tract, and clinical signs are not present. Fulminating septicemia, resulting in death with few clinical signs, can occur in immunosuppressed animals. There are a few reports of "circling" or "rolling" in mice associated with otitis media and interna caused by this organism.

Pathology. Gross and histopathologic lesions are nonspecific. Occasionally there is suppurative otitis media with extension into the inner ears and to the adjacent meninges or brain. Animals subjected to severe immunosuppression (e.g., whole-body irradiation or cyclophosphamide administration) can develop fulminant septicemia with hemorrhage and multifocal necrosis in multiple organs.

Diagnosis. Diagnosis of *P. aeruginosa* infection is made by isolation and identification of the organism and exclusion of other possible causes of disease. In animals that have been immunosuppressed, septicemia should be demonstrated by culture of the organism.

Control. Control is only necessary for immunosuppressed rodents. Cesarean derivation followed by maintenance under gnotobiotic conditions completely eliminates the organism. *P. aeruginosa* can be eliminated from animal facilities by rigorous sanitation measures coupled with acidification and/or hyperchlorination of the water.

Interference with Research. Indigenous infections are of little importance except when the research involves immunosuppressed animals. Mice and rats naturally infected with *P. aeruginosa* typically die earlier than do noninfected controls when exposed to lethal doses of whole-body irradiation, cyclophosphamide, cortisone, or other immunosuppressants.

Suggested Reading

Brownstein, D. G. 1978. Pathogenesis of *Pseudomonas aeruginosa* bacteremia in cyclophosphamide treated mice and potentiation of endogenous streptococcal virulence. J. Infect. Dis. 137:795-801.

Hammond, C. W. 1963. *Pseudomonas aeruginosa* infection and its effects on radiobiologic research. Lab. Anim. Care 13(1, Part 2):6-11.

Kohn, D. F., and W. F. MacKenzie. 1980. Inner ear disease characterized by rolling in C3H mice. J. Am. Vet. Med. Assoc. 177:815-817.

Matsumoto, T. 1982. Influence of *Escherichia coli*, *Klebsiella pneumoniae* and *Proteus vulgaris* on the mortality pattern of mice after lethal irradiation with γ rays. Lab. Anim. (London) 16:36-39.

McDougall, P. T., N. S. Wolf, W. A. Steinbach, and J. J. Trentin. 1967. Control of *Pseudomonas aeruginosa* in an experimental mouse colony. Lab. Anim. Care 17:204-215.

McPherson, C. W. 1963. Reduction of *Pseudomonas aeruginosa* and coliform bacteria in mouse drinking water following treatment with hydrochloric acid or chlorine. Lab. Anim. Care 13:737-744.

Reovirus-3

Agent. RNA virus. family Reoviridae, genus *Reovirus*.

Animals Affected. Mice, rats, hamsters, guinea pigs, and others.

Epizootiology. Reovirus-3 is prevalent in contemporary rodents. Transmission is by the fecal-oral route and probably by aerosol. Infected fomites may be important because reoviruses are relatively resistant to environmental conditions.

Clinical. Natural infections are subclinical.

Pathology. There are no pathologic changes associated with natural infections.

Diagnosis. The ELISA is the most sensitive method. The CF test is not reovirus-type specific, and the HAI test is prone to give false-positive results for reovirus-3. Virus isolations can be performed using L cells or embryonic kidney cells.

Transplantable tumors and cell lines can be screened for reoviruses by using tissue-culture methods or the MAP test.

Control. Cesarean derivation and barrier maintenance appear to be effective methods of control.

Interference with Research. Reovirus-3 is an occasional contaminant of and may interfere with research involving transplantable tumors and cell lines.

Suggested Reading

Hartley, J. W., W. P. Rowe, and R. J. Huebner. 1961. Recovery of reoviruses from wild and laboratory mice. Proc. Soc. Exp. Biol. Med. 108:390-395.

Nelson, J. B. 1964. Response of mice to reovirus type 3 in presence and absence of ascites tumour cells. Proc. Soc. Exp. Biol. Med. 116:1086-1089.

Rotavirus, Mouse

Agent. RNA virus, family Reoviridae, genus *Rotavirus*, group A.

Animals Affected. Mice.

Epizootiology. Mouse rotavirus is thought to be widely prevalent, but this has not been documented for contemporary mouse stocks. Transmission is by airborne infection in which contaminated dust and bedding from adjacent cages probably play key roles. Mice are most susceptible to infection from birth to about 17 days of age. Infected neonates shed high concentrations of virus in the feces from about 2 to 8-10 days postinfection. Transient viremia and viruria can occur. Mice infected after about 17 days of age shed lower concentrations of virus in the feces for 2-4 days. It is not known whether there is persistent infection or whether very low concentrations of virus are shed in the feces beyond these time points. There is no evidence for transplacental transmission.

Clinical. Susceptibility to disease is age dependent, being greatest from 4 to 17 days of age. Infection in adults is subclinical. Diarrhea during the first 2 weeks of life is the only consistent sign of disease. Watery, yellow stools usually begin around 48 hours postinfection and persist for about 1 week. Varying amounts of stool accumulate around the anus and base of the tail, soiling the coats of neonates and their dams. Affected neonates may appear lethargic and have distended abdomens. Usually there is no mortality.

Pathology. The infection progresses from the proximal to distal end of the intestine, involving the duodenum, jejunum, ileum, and colon. The virus selectively infects epithelial cytoplasm and destroys cells near the tips of the villi in the small intestine, resulting in mild villous atrophy. Cytoplasmic vacuoles of varying size occur in the mucosal epithelium.

Diagnosis. Diagnosis is made by serologic testing using the CF test, radioimmunoassay, IFA test, ELISA, or other methods; by demonstration of rotaviral antigen lesions in the small intestine; or by isolation of the virus using trypsinized primary monkey kidney cells.

Control. Cesarean derivation and exclusion by barrier maintenance are the traditional methods of control. However, it is more practical to use such methods as isolation and quarantine of individual breeding pairs with subsequent selection of seronegative progeny for breeding. The use of filter-top cage systems can be beneficial in controlling the transmission of infection among subpopulations housed in the same room.

Interference with Research. Infection can alter research results in studies using infant mice. Protein-calorie deprivation of nursing dams increases the severity of diarrhea, increases mortality, and reduces the weight gains of MRV-infected infant pups. Folic acid deficiency also increases the severity of diarrheal disease. Rotavirus-infected infant mice show increased mortality when challenged with enterotoxigenic *Escherichia coli*.

Suggested Reading

Eiden, J., H. M. Lederman, S. Vonderfecht, and R. Yolken. 1986. T-cell-deficient mice display normal recovery from experimental rotavirus infection. J. Virol. 57:706-708.

Eydelloth, R. S., S. L. Vonderfecht, J. F. Sheridan, L. D. Enders, and R. H. Yolken. 1984. Kinetics of viral replication and local and systematic immune responses in experimental rotavirus infection. J. Virol. 50:947-950.

Ferner, W. T., R. L. Miskuff, R. H. Yolken, and S. L. Vonderfecht. 1987. Comparison of methods for detection of serum antibody to murine rotavirus. J. Clin. Microbiol. 25:1364-1369.

Kraft, L. M. 1982. Viral diseases of the digestive system. Pp. 159-191 in The Mouse in Biomedical Research. Vol. II: Diseases, H. L. Foster, J. D. Small, and J. G. Fox, eds. New York: Academic Press.

Little, L. M., and J. A. Shadduck. 1982. Pathogenesis of rotavirus infection in mice. Infect. Immun. 38:755-763.

Riepenhoff-Talty, M., T. Dharakul, E. Kowalski, D. Sterman, and P. L. Orga. 1987. Rotavirus infection in mice: Pathogenesis and immunity. Adv. Exp. Med. Biol. 216B:1015-1023.

Salmonella enteritidis

Agent. Gram-negative bacterium, family Enterobacteriaceae, tribe Salmonellae. There are more than 1,500 serotypes of *Salmonella enteritidis*.

Animals Affected. Mice, rats, humans, and many others.

Epizootiology. Prevalence in laboratory rodents is unknown, but there is evidence that subclinical infections caused by weakly virulent strains are common in the United States. Infections are acquired primarily by ingestion, usually from contaminated food, water, or bedding. After initial colony infection, about 5% of the animals become asymptomatic, chronic carriers and shed organisms in the feces for many months.

Clinical. Most infections in mice are subclinical. In clinical outbreaks, nonspecific signs, including ruffled fur, hunched posture, reduced activity, weight loss, conjunctivitis, and variable mortality, follow an incubation period of 2-6 days. Diarrhea occurs in only about 20% of animals. Enzootically infected breeding colonies can have alternating periods of inapparent infection and overt clinical disease. Litter sizes and birth weights can be reduced. Clinical salmonellosis in rats is extremely rare. Virulence and dose of the organism and the host's age, strain, intestinal microflora, nutritional state, immune status, and intercurrent infections affect the expression of disease. C3H/HeN, C3H/St, C3H/Bi, CBA/Ca, BRVR, A/J, A/HeN, SWR/J, and DBA/2 mice are relatively resistant to *S. enteritidis*-induced disease, while BSVS, DBA/1, BALB/c, C57BL/6, C3H/HeJ, and CBA/N mice are relatively susceptible. Three distinct genetic loci affect susceptibility: *Ity* (immunity to *S. typhimurium*), *Lps* (lipopolysaccharide response), and *xid* (X-linked immune deficiency). Weanling mice are more susceptible than adults. Susceptibility to clinical disease is increased by food and water deprivation, nutritional iron deficiency in rats, nutritional iron overload in mice, pretreatment with sodium bicarbonate by gavage, and administration of morphine sulfate to slow gastrointestinal motility.

Pathology. Gross lesions are extremely variable, depending on the stage of disease. Animals that die of acute infection may have only hyperemia and congestion of visceral organs. Animals that survive a week or longer often appear emaciated, have hyperemia and thickening of the ileal and cecal walls, an empty or fluid-filled large intestine, multiple white or yellow foci in the liver, splenomegaly, enlarged mesenteric lymph nodes, and fibrinous exudate in the peritoneal cavity. Chronic carriers usually do not have gross lesions. The predominant microscopic lesions of salmonellosis are ileocecitis, mesenteric lymphadenitis, and multifocal inflammation in the liver and spleen, each varying in character depending on the stage of disease. There is multifocal to diffuse destruction of villous epithelium in the ileum, with blunting of villi and hyperplasia of crypt epithelium and purulent to pyogranulomatous inflammation in the lamina propria. Similar changes occur in the cecum, which is more severely affected in rats than in mice. Ulcerative cecitis, accompanied by severe pyogranulomatous inflammation in the lamina propria and purulent to chronic inflammation leading to atrophy and cyst formation in the paracecal lymph nodes, is characteristic of salmonellosis in rats. Multifocal purulent, pyogranulomatous, or granulomatous inflammation occurs in the mesenteric lymph nodes, liver, and spleen. So-called cell-fragment thrombi commonly occur in liver and spleen in which necrotic foci break into venous channels. Peritonitis is commonly caused by extension of the infection through the capsule of the liver, lymph nodes, or spleen. Cholangitis and cholecystitis are seen infrequently. Pyogranulomatous inflammation occasionally occurs in other organs such as the lungs.

Diagnosis. *S. enteritidis* is diagnosed by cultural isolation and serologic typing of isolates. It is usually best to culture the liver, spleen, intestine, feces, and blood. Characteristic histopathologic lesions help to rule out diseases with similar gross lesions (e.g., mousepox, Tyzzer's disease, and streptobacillosis in mice). Carriers must be detected by cultural isolation of the organism because there is no satisfactory serologic method for testing suspected carriers.

Control. Affected animals pose a zoonotic risk to personnel, are a source of infection for other animals, and are unsuitable for research. The usual method of control is to destroy the entire population and replace it with animals from a pathogen-free source. For extremely valuable stocks, cesarean derivations can be attempted. Prevention is by barrier maintenance, avoidance of *Salmonella*-contaminated food and water, and exclusion of infected animals and wild rodents from the animal facility. Regular monitoring is necessary.

Interference with Research. *Salmonella*-infected mice can have a nonspecific resistance to challenge with other intracellular parasites such as *Listeria monocytogenes*. Prior immunization of mice with viable *S. enteritidis* has resulted in the suppression of growth of transplantable tumors. Concurrent infections of *S. enteritidis* and *Plasmodium berghei* in mice have resulted in higher mortality than infection by either agent alone. *S. enteritidis* infection has caused reduced blood glucose and hepatic enzyme levels. Mice orally infected with *S. enteritidis* have reduced intestinal enzyme activities. Susceptibility to infection is increased by some experimental procedures.

Suggested Reading

Collins, F. M., and P. B. Carter. 1978. Growth of salmonellae in orally infected germfree mice. Infect. Immun. 21:41-47.

Ganaway, J. R. 1982. Bacterial and mycotic diseases of the digestive system. Pp. 1-20 in The Mouse in Biomedical Research. Vol. II: Diseases, H. L. Foster, J. D. Small, and J. G. Fox, eds. New York: Academic Press.

Casebolt, D. B., and T. R. Schoeb. 1988. An outbreak in mice of salmonellosis caused by *Salmonella enteritidis* serotype enteritidis. Lab. Anim. Sci. 38:190-192.

Farmer, J. J., III, B. R. Davis, F. W. Hickman-Brenner, A. McWhorter, G. P. Huntley-Carter, M. A. Asbury, C. Riddle, H. G. Wathen-Grady, C. Elias, G. R. Fanning, A. G. Steigerwalt, C. M. O'Hara, G. K. Morris, P. B. Smith, and D. J. Brenner. 1985. Biochemical identification of new species and biogroups of Enterobacteriaceae isolated from clinical specimens. J. Clin. Microbiol. 21:46-76.

Hsu, H. S. 1989. Pathogenesis and immunity in murine salmonellosis. Microbiol. Rev. 53:390-409.

Kirchner, B. K., L. W. Dixon, R. H. Lentsch, and J. E. Wagner. 1982. Recovery and pathogenicity of several salmonella species isolated from mice. Lab. Anim. Sci. 32:506-507.

Sendai Virus

Agent. RNA virus, family Paramyxoviridae, genus *Paramyxovirus*.

Animals Affected. Mice, rats, hamsters, and possibly guinea pigs.

Epizootiology. Sendai virus (SV) is extremely contagious. Prevalence is considered high in colonies of laboratory mice and rats worldwide. Natural infection occurs via the respiratory tract. Transmission is by direct contact and fomites and is highly efficient. Viral replication is thought to be limited to the respiratory tract and occurs for only about 1 week postinfection.

Clinical. Natural SV infection in rats is usually subclinical. In pregnant rats, the infection can cause fetal resorptions, retarded embryonic development, and mortality of neonates. Susceptibility to SV-induced disease in mice is dependent on the strain or stock. Those that are the most susceptible include 129/ReJ, 129/J, DBA/1J, DBA/2J, and S-*nu/nu* (Swiss carrying the mutation nude). Strains of intermediate susceptibility include A/HeJ, A/J, SWR/J, C57BL/10Sn, and BALB/c. The most resistant strains and stocks include SJL/J, RF/J, C57BL/6J, and S. The mode of inheritance and mechanisms of host resistance are poorly understood. Clinical disease caused by natural SV infection in mice falls into one of two patterns. Enzootic (subclinical) infection commonly occurs in breeding populations. Adults have active immunity due to prior infection, and newborn mice are passively protected by maternal antibody until around 4-8 weeks of age, when they become infected. Recovery is prompt and usually without morbidity or mortality. Epizootic (clinically apparent) infection occurs when a mouse population is first infected. Infection quickly spreads through the entire population. Signs are variable but may include chattering; mild respiratory distress; and prolonged gestation in adults, deaths in neonates and sucklings, and poor growth in weanling and young adults. Breeding colonies return to normal productivity in 2 months and thereafter maintain the enzootic pattern of infection. Epizootics of disease that exceed these general patterns in clinical severity should arouse suspicion of complication by other agent(s), particularly *Mycoplasma pulmonis* and cilia-associated respiratory bacillus.

Pathology. There are few gross lesions in uncomplicated SV infections. The lungs can appear focally reddened or atelectatic, and serous fluid can be visible in the pleural and pericardial cavities. The most severe lesions are seen in mice that are infected as sucklings or weanlings and in mice of the more susceptible strains. Severe necrotizing bronchitis and bronchiolitis often cause intense inflammatory injury to terminal bronchioles, resulting in scarring with severe distortion of the smaller airways and formation of polypoid outgrowths into bronchiole lumens. There is also pronounced hyperplasia of airway epithelium resulting in peribronchiolar adenomatous hyperplasia that can persist throughout life. In aged mice the air spaces in these lesions are often filled with mucus, large macrophages, and cellular debris. There can be large eosinophilic crystals in the air spaces, cytoplasm of the macrophages, and cells forming the adenomatoid structures. The terminal bronchioles of rats may be scarred and distorted but do not show the hyperplastic peribronchiolar changes seen in mice.

Athymic (*nu/nu*) mice develop chronic pneumonia similar to that in immunocompetent mice but have abundant intranuclear and intracytoplasmic inclusions in laryngeal, tracheal, bronchial, and bronchiolar epithelia, as well as in type I and II pneumocytes and alveolar macrophages. The virus persists for 10 weeks or longer. Athymic (*rnu/rnu*) rats also have increased susceptibility to SV and develop a similar chronic lung disease.

Diagnosis. The ELISA is the test of choice for routine serologic monitoring. The

ELISA also successfully detects anti-SV antibody in infected athymic (*nu/nu*) mice. An avidin-biotin-peroxidase-complex method has been used successfully for demonstrating SV antigen in histologic sections. Isolation of SV can be achieved by using BHK-21 or primary monkey kidney cell cultures or by inoculating the amniotic or allantoic sacs of 8- to 10-day-old embryonated hen's eggs. The MAP test can be used for determining contamination of transplantable tumors and other biologic materials.

Control. To prevent the introduction of infection, only animals known to be free of SV should be obtained, and the animals should be maintained under strict barrier conditions. In addition, all biologic materials, such as transplantable tumors, should be pretested and shown to be free of the virus. If SV infection is detected, prompt elimination of infected subpopulation(s) is essential to prevent spread of the infection to other rodents. A less effective alternative is to place the infected animals under strict quarantine, remove all young and pregnant females, suspend all breeding, and prevent the addition of other susceptible animals for a period of 6-8 weeks until the infection has run its course and the virus has been eliminated naturally. Cesarean derivation is effective but usually is not warranted. Vaccination might prove useful in some situations.

Interference with Research. Experimental SV infection alters the phagocytic function of pulmonary macrophages. Concurrent SV and *M. pulmonis* infections are synergistic in mice, causing disease of far greater severity than that caused by either agent alone. It has been reported (but not confirmed) that infected mice have deficiencies in T- and B-cell function that persist throughout life, but most of the evidence indicates that such deficiencies are transient, lasting only a few weeks. SV infection inhibits in vitro mitogenesis of lymphocytes, increases natural killer cell-mediated cytotoxicity, and increases cytotoxic lymphocyte responses after in vivo stimulation with SV-coated syngeneic cells. Isograft rejection is altered and the neoplastic response to respiratory carcinogens can be increased or decreased. Wound healing is delayed. Cyclophosphamide increases the clinical and pathologic severity of SV infection. SV infection in rats alters the mitogenic responses of T cells, reduces the severity of adjuvant arthritis, and decreases antibody response to sheep erythrocytes. SV infection alters host responses to transplantable tumors.

Suggested Reading

Brownstein, D. G. 1986. Sendai virus. Pp. 37-61 in Viral and Mycoplasmal Infections of Laboratory Rodents: Effects on Biomedical Research, P. N. Bhatt, R. O. Jacoby, H. S. Morse III, and A. E. New, eds. Orlando, Fla.: Academic Press.

Castleman, W. L., L. J. Brundage-Anguish, L. Kreitzer, and S. B. Neuenschwander. 1987. Pathogenesis of bronchiolitis and pneumonia induced in neonatal and weanling rats by parainfluenza (Sendai) virus. Am. J. Pathol. 129:277-286.

Hall, W. C., R. A. Lubet, C. J. Henry, and M. J. Collins, Jr. 1985. Sendai virus—Disease processes and research complications. Pp. 25-52 in Complications of Viral and Mycoplasmal Infections in Rodents to Toxicology Research and Testing, T. E. Hamm, Jr., ed. Washington, D.C.: Hemisphere.

Jakab, G. J. 1981. Interactions between Sendai virus and bacterial pathogens in the murine lung: A review. Lab. Anim. Sci. 31:170-177.

Parker, J. C., and C. B. Richter. 1982. Viral diseases of the respiratory system. Pp. 107-152 in The Mouse in Biomedical Research. Vol. II: Diseases, H. L. Foster, J. D. Small, and J. G. Fox, eds. New York: Academic Press.

Schoeb, T. R., K. C. Kervin, and J. R. Lindsey. 1985. Exacerbation of murine respiratory mycoplasmosis in gnotobiotic F344/N rats by Sendai virus infection. Vet. Pathol. 22:272-282.

Sialodacryoadenitis Virus

Agent. RNA virus, family Coronaviridae, genus *Coronavirus*.

Animals Affected. Rats. Mice are susceptible to experimental infection, but natural infection has not been reported for this species.

Epizootiology. Sialodacryoadenitis virus (SDAV) is one of the most common viruses in laboratory rats. It is highly contagious and transmitted by contact and aerosol. It is not transmitted vertically. LEW, WAG/Rij, and SHR rats are more susceptible than WI (Wistar), SD (Sprague-Dawley®), LE (Long Evans), and F344.

Clinical Signs. *Enzootic disease:* Adults are immune because of previous infection. Suckling rats have a mild, transient (1 week or less) conjunctivitis accompanied by blinking. Occasionally, exudate causes the eyelids to adhere together. Clinical signs usually disappear by weaning. *Epizootic disease:* Overt disease occurs in naive rat populations. There is a sudden high incidence of overt disease. Signs include cervical edema; sneezing; photophobia; serous to seropurulent, often porphyrin-stained nasal and ocular discharge; corneal ulceration; and keratoconus. There is high morbidity but no mortality. Most clinical signs disappear in a week, but the eyes might be more prominent than normal for 1-2 weeks because of inflammation of retroorbital tissues.

Pathology. SDAV has a positive tropism for serous or mixed serous-mucous tubuloalveolar glands. The submaxillary and parotid salivary, exorbital lacrimal, Harderian, and intraorbital lacrimal glands are the major target organs. There are also mild changes in the cervical lymph nodes, thymus, and respiratory tract. Characteristically, by 5 days postinfection, there is diffuse necrosis of alveolar and ductal epithelium in the salivary and lacrimal glands, and polymorphonuclear leukocytes quickly infiltrate the necrotic debris and interstitium accompanied by interstitial edema. The ductal epithelium is rapidly repaired, becoming hyperplastic and squamous in appearance by 10 days postinfection. Intranuclear inclusions are occasionally observed. Complete restoration of normal glandular architecture requires about 30 days. Eye lesions include interstitial keratitis, corneal ulceration, keratoconus, synechia, hypopyon, hyphema, and conjunctivitis. Sequelae of infection can include megaloglobus with lenticular and retinal degeneration. Thymic lesions are limited to focal necrosis of the cortex and medulla with some widening of the interlobular septae. Focal necrosis and lymphoid hyperplasia occur in cervical lymph nodes.

Diagnosis. The ELISA and the IFA test are more sensitive than the CF test. Presumptive diagnosis often can be based on characteristic histologic changes in Harderian, submaxillary, and parotid glands. Lesions can be bilateral or unilateral and are frequently found in animals with negative serologic tests for coronavirus antibody. The virus can be isolated by culture methods using primary rat kidney cells or by intracerebral inoculation of neonatal mice. The virus can be demonstrated in affected tissues by immunofluorescence for only about 7 days postinoculation.

Control. Control requires very strict adherence to preventative measures, including procurement only of rats known to be free of SDAV and adherence to strict barrier housing procedures. Prompt elimination of infected subpopulation(s) is essential to prevent spread of infection to other rodents. A less effective alternative is to place infected animals under strict quarantine, remove all young and pregnant females, suspend all breeding, and discontinue adding other susceptible animals for a period of 6-8 weeks until the infection has run its course and the virus has been eliminated naturally.

Interference with Research. The virus can seriously complicate studies involving the eyes, salivary glands, lacrimal glands, or respiratory tract. It is reported to reduce reproductive rate in breeding populations and slow growth rate of young rats. It inhibits phagocytosis and interleukin-1 production by pulmonary macrophages. SDAV infection exacerbates concurrent *Mycoplasma pulmonis* infection.

Suggested Reading

Bhatt, P. N., D. H. Percy, and A. M. Jonas. 1972. Characterization of the virus of sialodacryoadenitis of rats: A member of the coronavirus group. J. Infect. Dis. 126:123-130.

Boschert, K. R., T. R. Schoeb, D. B. Chandler, and D. L. Dillehay. 1988. Inhibition of phagocytosis and interleukin-1 production in pulmonary macrophages from rats with sialodacryoadenitis virus infection. J. Leukocyte Biol. 44:87-92.

Jacoby, R. O., P. N. Bhatt, and A. M. Jonas. 1975. Pathogenesis of sialodacryoadenitis in gnotobiotic rats. Vet. Pathol. 12:196-209.

Jacoby, R. O. 1986. Rat coronavirus. Pp. 625-638 in Viral and Mycoplasmal Infections of Laboratory Rodents: Effects on Biomedical Research, P. N. Bhatt, R. O. Jacoby, H. C. Morse III, and A. E. New, eds. Orlando, Fla.: Academic Press.

Schoeb, T. R., and J. R. Lindsey. 1987. Exacerbation of murine respiratory mycoplasmosis by sialodacryoadenitis virus infection in gnotobiotic F344 rats. Vet. Pathol. 24:392-399.

Wojcinski, Z. W., and D. H. Percy. 1986. Sialodacryoadenitis virus-associated lesions in the lower respiratory tract of rats. Vet. Pathol. 23:278-286.

Staphylococcus aureus

Agent. Gram-positive bacterium, family Micrococcaceae.

Animals Affected. Mice, rats, humans, and many others.

Epizootiology. *S. aureus* commonly colonizes the nasopharynx, lower digestive tract, fur, and skin. It is readily cultured from cages, room surfaces, and personnel. The epizootiology of pathogenic types in animals is poorly understood. Human carriers might be an important source of infection for rodent colonies and vice versa. Many rodent colonies have infection without overt disease. Pathogenesis is probably dependent on many factors, including phage type(s), traumatic injuries of skin or mucosal surfaces, host factors, and sanitation.

Clinical. *Rats:* Ulcerative dermatitis with intensely pruritic, moist eczematous lesions (usually 1-2 cm in diameter) occurs on the lateral surfaces of the shoulders and neck. Lesions appear to be initiated or aggravated by scratching. *Mice:* Ulcerative dermatitis with moist eczematous lesions occurs on the face, neck, ears, and forelegs. Multiple abscesses and botryomycotic granulomas can develop in deeper tissues of the face, including the orbital tissues, facial muscles, peridontium, and mandibles. Purulent lesions of varying size occur commonly around the eyes and on the face of athymic (*nu/nu*) mice. Abscesses also occur in preputial glands, which become firm and enlarged to a few millimeters in diameter. The highest incidence of preputial gland abscesses is in strain C3H/HeN. In strain C57BL/6N, the organism has been associated with a syndrome involving self-mutilation of the penis.

Pathology. Suppurative inflammation is a hallmark of tissue invasion by *S. aureus*. In ulcerative dermatitis, there is destruction of the epidermis, and the underlying dermis contains pustules; abscesses; and, eventually, chronic or granulomatous inflammation. Large numbers of organisms are usually present and can be demonstrated readily in Gram-stained sections or imprints. Preputial abscesses in C3H/HeN mice apparently are caused by ascending infection from the ducts of the preputial glands.

Diagnosis. Diagnosis depends on the isolation and identification of *S. aureus* and exclusion of other agents (e.g., dermatophytes and mites) as possible causes of the lesions.

Control. The best methods of control are improved sanitation, frequent sterilization of cages and other equipment, elimination of equipment that could cause skin injury, and reduction in the number of animals per cage.

Interference with Research. *S. aureus* alters host immune responses (e.g., activates suppressor B cells). Rats maintained on prolonged immunosuppression with corticosteroids can develop *S. aureus*-induced renal abscesses.

Suggested Reading

Clarke, M. C., R. J. Taylor, G. A. Hall, and P. W. Jones. 1978. The occurrence of facial and mandibular abscesses associated with *Staphylococcus aureus*. Lab. Anim. (London) 12:121-123.

Easmon, C. S. F., and A. A. Glynn. 1979. The cellular control of delayed hypersensitivity to *Staphylococcus aureus* in mice. Immunology 38:103-108.

Hong, C. C., and R. D. Ediger. 1978. Self-mutilation of the penis in C57BL/6N mice. Lab. Anim. (London) 12:55-57.

Hong, C. C., and R. D. Ediger. 1978. Preputial gland abscess in mice. Lab. Anim. Sci. 28:153-156.
Fox, J. G., S. M. Niemi, J. C. Murphy, and F. W. Quimby. 1977. Ulcerative dermatitis in the rat. Lab. Anim. Sci. 27:671-678.
Wagner, J. E., D. R. Owens, M. C. LaRegina, and G. A. Vogler. 1977. Self-trauma and *Staphylococcus aureus* in ulcerative dermatitis in rats. J. Am. Vet. Med. Assoc. 171:839-841.

Streptobacillus moniliformis

Agent. Gram-negative bacterium; taxonomy uncertain.

Animals Affected. Rats. Mice, guinea pigs, humans, and others can contract the infection from rats.

Epizootiology. *S. moniliformis* is a commensal of the nasopharynx in wild and conventionally reared laboratory rats. It has not been reported in cesarean-derived, barrier-maintained rats or mice. Epizootic disease is most likely to occur in mice housed near infected rats. Transmission is by rat bites, aerosols, and fomites.

Clinical. In rats the infection is subclinical. The organism can be isolated from the nasopharynx, middle ear, respiratory tract, and subcutaneous abscesses. In mice, early signs include conjunctivitis, photophobia, cyanosis, diarrhea, anemia, hemoglobinuria, emaciation, and high mortality. Septicemia clears in survivors in a few weeks, but infection persists around the joints for about 6 months. During this chronic phase of infection, there can be diffuse swelling and reddening of limbs or tail, with the development of chronic arthritis, deformity and ankylosis, or amputation (ectromelia). Spinal lesions, accompanied by posterior paralysis, kyphosis, and priapism, can occur. Pregnant females can abort or produce stillborn young.

Pathology. In mice, early lesions are associated with septicemia, including focal necrosis of the spleen and liver, splenomegaly, and lymphadenopathy. Subsequent lesions are primarily those of chronic arthritis in various stages of development and severity.

Diagnosis. Diagnosis depends on cultural isolation and identification of the organism. Differential diagnosis should distinguish between *S. moniliformis*-induced disease and mousepox or bacterial septicemias caused by other organisms (e.g., *Corynebacterium kutscheri, Salmonella enteritidis*).

Control. Cesarean derivation, barrier maintenance, and regular monitoring for rodent pathogens by a comprehensive health surveillance program are the best methods of control. Mice should not be housed in the same room as rats that have not been monitored for *S. moniliformis* infection.

Interference with Research. There have been no reports of interference with the results of research in which rats were used. *S. moniliformis* can cause high mortality in mice and is a serious zoonotic infection in humans.*

---

*In humans the incubation period is usually 3-10 days, followed by the abrupt onset of fever, chills, vomiting, headache, and myalgia. There is a maculopapular rash that is most pronounced on the extremities. Arthritis occurs in about two-thirds of cases, and other complications such as endocarditis and focal abscesses occur in some untreated cases.

Suggested Reading

Anderson, L. C., S. L. Leary, and P. J. Manning. 1983. Rat-bite fever in animal research laboratory personnel. Lab. Anim. Sci. 33:292-294.

Lerner, E. M., II, and L. Sokoloff. 1959. The pathogenesis of bone and joint infections produced by *Streptobacillus moniliformis*. Arch. Pathol. 67:364-372.

Savage, N. 1984. Genus *Streptobacillus*. Pp. 598-600 in Bergey's Manual of Systematic Bacteriology, vol. I, N. R. Kreig and J. G. Holt, eds. Baltimore: Williams & Wilkins.

Savage, N. L., G. N. Joiner, and D. W. Florey. 1981. Clincial, microbiological, and histological manifestations of *Streptobacillus moniliformis*-induced arthritis in mice. Infect. Immun. 34:605-609.

Weisbroth, S. H. 1979. Bacterial and mycotic diseases. Pp. 191-241 in The Laboratory Rat. Vol. I: Biology and Diseases, H. J. Baker, J. R. Lindsey, and S. H. Weisbroth, eds. New York: Academic Press.

Streptococcus pneumoniae

Agent. Gram-positive bacterium, family Streptoccaceae.

Animals Affected. Only subclinical infection has been reported in mice. Disease has been reported occasionally in rats, guinea pigs, and monkeys. Humans are the main natural hosts.

Epizootiology. Transmission is mainly by aerosol.

Clinical Signs. Signs include dyspnea, weight loss, hunched posture, snuffling respiratory sounds, and abdominal breathing. Clinical onset can be sudden, and young rats are affected most often.

Pathology. Predominant lesions in rats are suppurative rhinitis and otitis media. The disease often extends into distal airways, causing acute tracheitis and fibrinous lobar pneumonia, and into organs adjacent to the lungs, causing fibrinous pleuritis or empyema, fibrinous pericarditis, and/or acute mediastinitis. Lesions associated with severe bacteremia include suppurative arthritis, meningitis, hepatitis, splenitis, peritonitis, and orchitis. Splenic and testicular infarcts can occur. Abdominal lesions are frequently the primary cause of death.

Diagnosis. Diagnosis is made by isolating the organism from sites with characteristic lesions and excluding other possible causes and contributors to the disease.

Control. Cesarean derivation and barrier maintenance are extremely effective methods of control. It might be helpful for personnel to wear masks because of the high prevalence of the infection in humans.

Interference with Research. *S. pneumoniae*-induced septicemia alters hepatic metabolism, serum biochemistries, and thyroid function. Studies involving the rat respiratory tract can be jeopardized.

Suggested Reading

Fallon, M. T., M. K. Reinhard, B. M. Gray, T. W. Davis, and J. R. Lindsey. 1988. Inapparent

Streptococcus pneumoniae type 35 infections in commercial rats and mice. Lab Anim. Sci. 38:129-132.

Harding, B., F. Kenny, F. Given, B. Murphy, and S. Lavelle. 1987. Autotransplantation of splenic tissue after splenectomy in rats offers partial protection against intravenous pneumococcal challenge. Eur. Surg. Res. 19:135-139.

Johanson, W. G., Jr., S. J. Jay, and A. K. Pierce. 1974. Bacterial growth *in vivo*. An important determinant of the pulmonary clearance of *Diplococcus pneumoniae* in rats. J. Clin. Invest. 53:1320-1325.

Quie, P. G., G. S. Giebink, and J. A. Winkelstein, eds. 1981. The pneumococcus. Rev. Infect. Dis. 3:183-395.

Weisbroth, S. H. 1979. Bacterial and mycotic diseases. Pp. 193-241 in The Laboratory Rat. Vol. I: Biology and Diseases, H. J. Baker, J. R. Lindsey, and S. H. Weisbroth, eds. New York: Academic Press.

Wood, W. B., Jr., and M. R. Smith. 1950. Host-parasite relationships in experimental pneumonia due to pneumococcus type III. J. Exp. Med. 92:85-100.

Theiler's Murine Encephalomyelitis Virus

Agent. RNA virus, family Picornaviridae, genus *Enterovirus*.

Animals Affected. Laboratory mice and rats. The infection has not been reported in wild mice or rats.

Epizootiology. The prevalence of Theiler's murine encephalomyelitis virus (TMEV) in mice is generally thought to be very low in barrier breeding colonies in the United States. Prevalence in laboratory rats is unknown. In naturally infected mice, there is a low titer of virus in intestinal mucosa; intestinal contents; feces; and, less frequently, mesenteric lymph nodes. Virus shedding in the feces has been documented to occur as long as 154 days postinfection. Transmission is by the fecal-oral route. Transplacental infection does not occur.

Clinical. Natural infections in mice are usually inapparent and, presumably, are caused by less virulent, wild-type strains of TMEV resembling Theiler's original (TO) strain of virus. Clinical disease appears at a rate of only 1 in 4,000-10,000 infected animals. Affected mice may have flaccid paralysis of one or both rear legs. There is little or no mortality. Signs of infection with a highly virulent strain of TMEV in one natural outbreak included circling, rolling, hyperexcitability, convulsions, tremors, weakness or flaccid paralysis of the hind legs, and high mortality. Cinical signs of infection with the MHG strain of TMEV in rats included circling, incoordination, tremors, and torticollis.

Pathology. Natural disease in mice results from the rare occurrence of viremia, i.e., the dissemination of virus from the intestine to the spinal cord and brain. This occurs most frequently around 6-10 weeks of age. The predominant lesion is poliomyelitis, with necrosis and neuronophagia of anterior horn cells and nonsuppurative inflammation composed primarily of lymphocytes. Little if any secondary demyelination is seen in the natural disease. TMEV can be isolated from the lesions for at least 1 year.

Diagnosis. The ELISA is the method of choice for serologic screening. If the HAI test with GDVII antigen and human type O erythrocytes is used, it is essential to perform the test at 4°C to avoid false-positive results. The CF and serum neutralization tests might also be useful for some purposes, such comparison of the cross-reactivity of TMEV strains. The MAP test can be used for screening biologic materials. Definitive diagnosis usually is made by isolating the virus from the spinal cords or brains of mice with clinical disease, but it is also possible to isolate the virus from the intestinal contents of mice with asymptomatic infections.

Control. The most practical method of control is to obtain mice from breeding populations that have been shown serologically to be free of the infection, followed by barrier maintenance and regular testing to reconfirm TMEV-free status. TMEV infection has been eliminated from valuable mouse stocks by foster nursing infant mice on TMEV-free mice or rats. Cesarean derivation is also effective but usually is not warranted.

Interference with Research. Indigenous TMEV infections occasionally interfere with studies of other unrelated viruses in mice.

Suggested Reading

Brownstein, D., P. Bhatt, R. Ardito, F. Paturzo, and E. Johnson. 1989. Duration and patterns of transmission of Theiler's mouse encephalomyelitis virus infection. Lab. Anim. Sci. 39:299-301.

Downs, W. G. 1982. Mouse encephalomyelitis virus. Pp. 341-352 in The Mouse in Biomedical Research. Vol. II: Diseases, H. L. Foster, J. D. Small, and J. G. Fox, eds. New York: Academic Press.

McConnell, S. J., D. L. Huxsoll, F. M. Garner, R. O. Spertzel, A. R. Warner, Jr., and R. H. Yager. 1964. Isolation and characterization of a neurotropic agent (MHG virus) from adult rats. Proc. Soc. Exp. Biol. Med. 115:362-367.

Theiler, M., and S. Gard. 1940. Encephalomyelitis of mice. III. Epidemiology. J. Exp. Med. 72:79-90.

Thompson, R., V. M. Harrison, and F. P. Myers. 1951. A spontaneous epizootic of mouse encephalomyelitis. Proc. Soc. Exp. Biol. Med. 77:262-266.

von Magnus, H., and P. von Magnus. 1948. Breeding of a colony of white mice free of encephalomyelitis virus. Acta Pathol. Microbiol. Scand. 26:175-177.

Thymic Virus, Mouse

Agent. Considered a herpesvirus because of its ultrastructural features and properties of heat and ether lability.

Animals Affected. Wild and laboratory mice.

Epizootiology. Wild mice apparently serve as reservoir hosts. Prevalence in laboratory mice is unknown; however, limited data suggest natural infections might be common. Mouse thymic virus (MTV) apparently occurs as a persistent subclinical infection in salivary glands, with virus shed in saliva. Horizontal transmission is most important, but vertical transmission cannot be ruled out.

Clinical Signs. Natural infections are subclinical.

Pathology. There are no pathologic changes associated with natural infections.

Diagnosis. The ELISA is the most sensitive serologic test, although the IFA and CF tests are also in use. However, these tests are only useful for screening adult mice; neonatally infected mice do not produce serum antibody. To detect MTV infection in neonatal mice, it is necessary to inoculate pathogen-free neonatal mice with salivary gland homogenate, saliva, or other material from the neonatal mice to be tested. If the test mice are infected with MTV, histologic examination of the thymuses, lymph nodes, and spleens of the pathogen-free mice 10-14 days postinoculation will disclose lymphoid necrosis and intranuclear inclusions. Virus isolation is impossible because no cell culture system is known to support its growth.

Control. No data are available.

Interference with Research. MTV infection might complicate experiments involving the passage of tissues in neonatal mice.

Suggested Reading

Cross, S. S. 1973. Development of Bioassays and Studies on the Biology of Mouse Thymic Virus. Ph.D. Dissertation. Washington, D.C.: George Washington University.

Cross, S. S., J. C. Parker, W. P. Rowe, and M. L. Robbins. 1979. Biology of mouse thymic virus, a herpesvirus of mice, and the antigenic relationship to mouse cytomegalovirus. Infect. Immun. 26:1186-1195.

Lussier, G., D. Guenette, W. R. Shek, and J. P. Descoteaux. 1988. Detection of antibodies to mouse thymic virus by enzyme-linked immunosorbent assay. Can. J. Vet. Res. 52:236-238.

Morse, S. S. 1987. Mouse thymic necrosis virus: A novel murine lymphotropic agent. Lab. Anim. Sci. 37:717-725.

Morse, S. S., and J. E. Valinsky. 1989. Mouse thymic virus (MTLV). A mammalian herpesvirus cytolytic for $CD4^+$ ($L3T4^+$) T lymphocytes. J. Exp. Med. 169:591-596.

Rowe, W. P., and W. I. Capps. 1961. A new mouse virus causing necrosis of the thymus in newborn mice. J. Exp. Med. 113:831-844.

DERMATOPHYTES

Trichophyton spp. and *Microsporum* spp.

Agents. Fungi.

Animals Affected. Mice, rats, humans, and numerous other animals.

Epizootiology. Dermatophytes have not been reported in cesarean-derived, barrier-maintained rodent stocks. Other animals are probably the major reservoirs of infection for mice and rats. The organisms are parasites of keratin, i.e., hair and superficial layers of skin.

Clinical. Infections are rare, and when they do occur, they are usually subclinical. Clinical disease has been seen more frequently in mice than in rats.

Lesions consist of irregularly defined areas of alopecia with a scaly to crusty appearance and occasional pustules at the edges. Lesions most commonly occur on the head near the mouth and eyes, but they can be found anywhere on the body.

Pathology. Uncomplicated lesions are very subtle; microscopic examination reveals only thickening of the stratum corneum in sections stained with hematoxylin and eosin. Special stains such as periodic acid-Schiff or Gridley's fungus stain are valuable in demonstrating the organisms.

Diagnosis. If infection is suspected in asymptomatic animals, several animals should be held over opened plates of culture medium while the fur is brushed. The plates should then be cultured for dermatophytes. In clinical cases, the hair should be plucked or skin scrapings should be taken from the periphery of the lesions and mounted onto slides in 10% potassium hydroxide for visualization of hyphae and endospores. Definitive diagnosis is dependent on culture and identification of the organisms by using Sabouraud's or other dermatophyte mediums.

Control. Where feasible, infected stocks should be destroyed and replaced by dermatophyte-free stock after thorough sterilization and disinfection of the facilities and equipment. Treatment of affected animals is not recommended. For prevention of infection, barrier maintenance appears to be effective. Rodents should be housed well away from laboratory animal species known to be more frequently infected (e.g., cats and dogs). Dermatophyte infections of mice and rats are rare in contemporary stocks and, therefore, do not represent important zoonoses.

Suggested Reading

Balsari, A., C. Bianchi, A. Cocilova, I. Dragoni, G. Poli, and W. Ponti. 1981. Dermatophytes in clinically healthy laboratory animals. Lab. Anim. 15:75-77.

Fox, J. G., and J. B. Brayton. 1982. Mycoses. Pp. 411-413 in The Mouse in Biomedical Reseach. Vol. II: Diseases, H. L. Foster, J. D. Small, and J. G. Fox, eds. New York: Academic Press.

Kunstyr, I. 1980. Laboratory animals as a possible source of dermatophytic infections in man. Zbl. Bakt. Med. Mycol. 8:361-367.

Weisbroth, S. H. 1979. Dermatomycosis (*Trichophyton mentagrophytes*). Pp. 228-229 in The Laboratory Rat. Vol. I: Biology and Diseases, H. J. Baker, J. R. Lindsey, and S. H. Weisbroth, eds. New York: Academic Press.

Williford, C. B., and J. E. Wagner. 1982. Mycotic diseases. Pp. 65-68 in The Mouse in Biomedical Research. Vol. II: Diseases, H. L. Foster, J. D. Small, and J. G. Fox, eds. New York: Academic Press.

COMMON ECTOPARASITES

Myobia musculi

Agent. Fur mite, order Acarina.

Life Cycle. *Myobia musculi* has egg, larval, nymphal, and adult stages. Eggs are oval, about 200 µm long, and usually found either attached to the base of hairs or inside mature females. Eggs hatch in about 7 days, and completion of the entire life cycle requires about 23 days.

Animals Affected. Mice and rarely rats and other laboratory rodents.

Epizootiology. Mites can be seen anywhere on the body but are most numerous alongside the hair bases in the more densely furred parts of the body (i.e., over the head and back). Transmission is by direct contact. The dynamics of mite populations on a host are very complex and are influenced by factors that include grooming, strain susceptibility, and host immune responses. Athymic (*nu/nu*) and other furless mice are not susceptible to infestation.

Clinical. The general appearance of infested mice is not directly related to the size of the mite population. Infestations are commonly subclinical. Clinical signs include scruffiness, pruritis, patchy alopecia, self-trauma, ulceration of the skin, and pyoderma. Close inspection often reveals bran-like hyperkeratotic debris and mites on the skin around the base of the hairs.

Pathology. Mice of the C57BL strains and their congeneic sublines are particularly susceptible to severe *M. musculi*-related skin disease. Lesions vary from mild to severe. Initially there is mild hyperkeratosis, but this often progresses to severe hyperkeratosis with fine bran-like material on the skin over virtually all of the body but particularly abundant over the dorsum, head, and shoulders. In more advanced cases, there is patchy alopecia and chronic ulcerative dermatitis most frequently distributed asymmetrically in the shoulder and neck regions. Secondary bacterial infection commonly leads to suppurative and granulomatous inflammation. Hyperplasia of regional lymph nodes, splenic lymphoid hyperplasia, and increased serum immunoglobulins are common.

Diagnosis. Diagnosis requires demonstration and identification of mites, while excluding other causes of dermatitis such as fungi (ringworm) or *Staphylococcus aureus*. Mites can be demonstrated by using a stereoscopic microscope or hand lens to examine the pelage, particularly over the back and head. Alternatively, mice can be killed and placed either on black paper and left at room temperature or in tape-sealed Petri dishes and refrigerated for an hour. As the body cools, the mites leave it and can be collected from the paper or Petri dish. The mites are mounted under a coverslip on glass slides with immersion oil and identified microscopically on the basis of anatomic features.

Control. Cesarean derivation and barrier maintenance are the most effective methods for eradication of mite infestations. Insecticides can be used, but they may alter experimental results.

Interference with Research. Behavioral patterns are likely to be altered by hypersensitivity to these mites. Secondary amyloidosis caused by chronic infestation can interfere with research results.

Suggested Reading

Flynn, R. J. 1973. Parasites of Laboratory Animals. Ames, Iowa: Iowa State University Press. 884 pp.

Friedman, S., and S. H. Weisbroth. 1977. The parasitic ecology of the rodent mite, *Myobia musculi*. IV. Life cycle. Lab. Anim. Sci. 27:34-37.

Galton, M. 1963. Myobic mange in the mouse leading to skin ulceration and amyloidosis. Am. J. Pathol. 43:855-865.

Weisbroth, S. H. 1982. Arthropods. Pp. 385-402 in The Mouse in Biomedical Research. Vol. II: Diseases, H. L. Foster, J. D. Small, and J. G. Fox, eds. New York: Academic Press.

Weisbroth, S. H., S. Friedman, and S. Scher. 1976. The parasitic ecology of the rodent mite *Myobia musculi*. III. Lesions in certain host strains. Lab. Anim. Sci. 26:725-735.

Myocoptes musculinus and *Radfordia affinis*

Agent. Mites, order Acarina.

Pathology. *M. musculinus* causes lesions similar to, but usually milder than, those caused by *Myobia musculi*. *R. affinis* is not a significant pathogen.

Interference with Research. Mite infestations due to *M. musculinus* have been reported to reduce the contact sensitivity of mice to oxazolone.

Suggested Reading

Flynn, R. J. 1973. Parasites of Laboratory Animals. Ames, Iowa: Iowa State University Press. 884 pp.

Laltoo, H., and L. S. Kind. 1979. Reduction in contact sensitivity reactions to oxazolone in mite-infested mice. Infect. Immun. 26:30-35.

Weisbroth, S. H. 1982. Arthropods. Pp. 385-402 in The Mouse in Biomedical Research. Vol. II: Diseases, H. L. Foster, J. D. Small, and J. G. Fox, eds. New York: Academic Press.

Other Ectoparasites

For more in-depth coverage or information on less common ectoparasites of mice and rats, consult comprehensive reference works on the subject.

Suggested Reading

Flynn, R. J. 1973. Parasites of Laboratory Animals. Ames, Iowa: Iowa State University Press. 884 pp.

Hsu, C.-K. 1979. Parasitic diseases. Pp. 307-331 in The Laboratory Rat. Vol. I: Biology and Diseases, H. J. Baker, J. R. Lindsey, and S. H. Weisbroth, eds. New York: Academic Press.

Owen, D. 1972. Common Parasites of Laboratory Rodents and Lagomorphs. MRC Laboratory Animals Centre Handbook No. 1. London: Medical Research Council. 64 pp.

Pratt, H. D., and J. S. Wiseman. 1962. Fleas of Public Health Importance and Their Control. Insect Control Series: Part VIII. PHS Publ. No. 772. Washington, D.C.: U.S. Department of Health, Education, and Welfare.

Weisbroth, S. H. 1982. Arthropods. Pp. 385-402 in The Mouse in Biomedical Research. Vol. II: Diseases, H. L. Foster, J. D. Small, and J. G. Fox, eds. New York: Academic Press.

ENDOPARASITES

Aspicularis tetraptera (Mouse Pinworm)

Agent. Roundworm, order Ascarida, suborder Oxyurina.

Life Cycle. Direct, requires 23-25 days. The adults reside in the colon. Females lay their eggs in the colon, and the eggs subsequently leave the host on fecal pellets. The eggs become infective after 6-7 days at room temperature. Transmission occurs when the infective eggs are ingested by another host. The eggs hatch in the colon, where the larvae develop to maturity, and the cycle begins again.

Animals Affected. Mice, rats (rarely), and wild rodents.

Epizootiology. *A. tetraptera* inhabits and lays eggs in the colon. The eggs survive for weeks in animal room environments.

Clinical. Infections are subclinical.

Pathology. *A. tetraptera* is not considered pathogenic.

Diagnosis. Diagnosis is made by demonstration of distinctive eggs by fecal flotation (the cellophane tape method is of no value) and by demonstration and identification of the adult worms in the colon at necropsy.

Control. Cesarean derivation and barrier maintenance are effective. Infection can be controlled to some extent by using hygienic methods, such as frequent cage and room sanitization. Cage-to-cage transmission can be prevented by using filter-top cages. Several anthelminthics are effective in eliminating a high percentage of adult worms, but many are inefficient in clearing immature worms or eggs.

Interference with Research. See *Syphacia obvelata* (p. 66).

Suggested Reading

Flynn, R. J. 1973. Nematodes. Pp. 203-320 in Parasites of Laboratory Animals. Ames, Iowa: Iowa State University Press.

Hsu, C.-K. 1979. Parasitic diseases. Pp. 307-331 in The Laboratory Rat. Vol. I: Biology and Diseases, H. J. Baker, J. R. Lindsey, and S. H. Weisbroth, eds. New York: Academic Press.

Wescott, R. B. 1982. Helminths. Pp. 374-384 in The Mouse in Biomedical Research. Vol. II: Diseases, H. L. Foster, J. D. Small, and J. G. Fox, eds. New York: Academic Press.

Entamoeba muris

Agent. Protozoan, order Amoebida, family Entamoebidae.

Life Cycle. Direct. Trophozoites, which inhabit the cecum and colon, form cysts that are passed in the feces. Transmission is by ingestion of cysts.

Animals Affected. Mice, rats, hamsters, and other rodent species.

Epizootiology. Trophozoites are most commonly found at the interface between the fecal stream and the intestinal epithelium in the cecum and colon. Cysts are resistant to environmental conditions.

Clinical. Infection is subclinical.

Pathology. The organism is nonpathogenic.

Diagnosis. Diagnosis is made by demonstrating cysts in the feces or trophozoites in wet mounts of intestinal contents from the cecum or colon. In sections stained by hematoxylin and eosin, the trophozoites usually have a distinct magenta-stained nucleus and violet-stained cytoplasm that can appear vacuolated. The outer cell membrane of the trophozoites is usually distinctly visible.

Control. *Entamoeba muris* can be eliminated by cesarean derivation and barrier maintenance; however, infection with this agent is generally considered inconsequential, and control measures are usually not necessary.

Interference with Research. There have been no reports of interference with research results.

Suggested Reading

Levine, N. D. 1961. Protozoan Parasites of Domestic Animals and of Man. Minneapolis, Minn.: Burgess. 412 pp.

Levine, N. D. 1974. Diseases of laboratory animals—parasitic. Pp. 209-327 in CRC Handbook of Laboratory Animal Science, vol. II, E. C. Melby and N. H. Altman, eds. Cleveland: CRC Press.

Giardia muris

Agent. Flagellated protozoan, order Diplomonadida, family Hexamitidae, subfamily Giardinae.

Life Cycle. Direct. Trophozoites reproduce by longitudinal fission and form cysts that are passed in the feces. Transmission is by ingestion of cysts. The minimal infectious dose for a mouse is approximately 10 cysts.

Animals Affected. Mice, rats, hamsters, humans, and many other species.

Epizootiology. Trophozoites colonize the proximal one-fourth of the small intestine, where they are found mainly adhering to columnar cells of the villi and free in the adjacent mucous layer. The number of trophozoites in the small intestine correlates directly with the number of cysts in the large intestine and feces. Cysts are resistant to most environmental conditions but are inactivated by treatment with a 2.5% phenol solution and by temperatures above 50°C.

Clinical. Infections in mice and rats are usually subclinical but can cause reduced weight gain, rough hair coats, and enlarged abdomens. Infection may be associated with morbidity and mortality in athymic (*nu/nu*) and other immunocompromised mice.

Pathology. Pathogenesis has been studied most extensively in mice. The acute phase of infection involves the proliferation of trophozoites in the small intestine, with the peak period of cyst release occurring during the second week of infection. In the elimination phase, cysts released in the feces are reduced to undetectable levels. Resistant strains, including DBA/2, B10.A, C57BL/6, BALB/c, and SJL/J, eliminate the infection in 5 weeks. Susceptible strains and stocks, including C3H/He and A/J and outbred Crl:ICR (CD®-1), require 10 weeks to eliminate the infection. Highly susceptible athymic (*nu/nu*) mice have prolonged infections. Resistance during the acute phase of infection is thought to be controlled by several genes not linked to the *H-2* locus, while resistance during the elimination phase is inherited as a dominant trait. Protective immunity is probably dependent on both antibody- and cell-mediated mechanisms. The milk of immune mice contains both IgA and IgG antibodies against *Giardia muris* and conveys passive protection. In uncomplicated *G. muris* infection, morphological changes in the small intestine are usually minimal. The villus to crypt ratio may be reduced, and variable numbers of lymphocytes may be present.

Diagnosis. Infection by other possible primary or contributing pathogens must be excluded. The organism is diagnosed histologically by identifying characteristic "monkey-faced" trophozoites in sections of the small intestine. Trophozoites also can be recognized in wet mounts of intestinal contents by their characteristic shape and their rolling and tumbling motion. Cysts can be demonstrated in wet mounts of feces.

Control. The most practical approach to controlling infection is to procure rodents from breeding populations shown by health surveillance testing to be free of *G. muris* and to maintain them in a barrier facility. Cesarean derivation is required to eliminate the parasite from infected stocks. Metronidazole can be used for treatment of infected animals but does not completely eradicate infection.

Interference with Research. Infection with *G. muris* can increase the severity and mortality of wasting syndrome (presumably caused by mouse hepatitis virus) in athymic (*nu/nu*) mice. The organism causes a transient reduction in immunoresponsiveness of mice to sheep erythrocytes during the second and third weeks of infection. It also alters intestinal fluid accumulation and mucosal immune responses caused by cholera toxin in mice.

Suggested Reading

Belosevic, M., G. M. Faubert, E. Skamene, and J. D. MacLean. 1984. Susceptibility and resistance of inbred mice to *Giardia muris*. Infect. Immun. 44:282-286.

Boorman, G. A., P. H. C. Lina, C. Zurcher, and H. T. M. Nieuwerkerk. 1973. *Hexamita* and *Giardia* as a cause of mortality in congenitally thymus-less (nude) mice. Clin. Exp. Immunol. 15:623-637.

Brett, S. J. 1983. Immunodepression in *Giardia muris* and *Spironucleus muris* infections in mice. Parasitology 87:507-515.

Brett, S. J., and F. E. G. Cox. 1982. Immunological aspects of *Giardia muris* and *Spironucleus muris* infections in inbred and outbred strains of laboratory mice: A comparative study. Parasitology 85:85-99.

Hsu, C.-K. 1982. Protozoa. Pp. 359-372 in The Mouse in Biomedical Research. Vol. II: Diseases, H. L. Foster, J. D. Small, and J. G. Fox, eds. New York: Academic Press.

Stevens, D. P. 1982. Giardiasis: Immunity, immunopathology and immunodiagnosis. Pp. 192-203 in Immunology of Parasitic Infections, 2nd ed., S. Cohen and K. S. Warren, eds. Oxford: Blackwell Scientific.

Hymenolepis nana

Agent. Tapeworm, order Cyclophyllidea, family Hymenolepidae.

Life Cycle. Direct or indirect. The life cycle includes adult, egg (with embryo or oncosphere), and larval (cercocystis) stages. In direct transmission, eggs hatch in the small intestine. Larvae penetrate and develop as cercocystis in the intestinal villi, then return to the lumen to become mature adults. The cycle requires only 1-16 days. In indirect transmission, the eggs are ingested by an arthropod intermediate host such as a flour beetle, and the cercocystis develops in the intestine of the beetle. The intermediate host is eaten by the definitive host, and adult *H. nana* develop in the lumen of the small intestine. The entire life cycle by indirect transmission requires 20-30 days.

Animals Affected. Mice, rats, hamsters, other rodents, nonhuman primates, and humans.

Epizootiology. Weanling and young adult rodents are most frequently infected. The duration of infection by adult worms in the small intestine is usually only a few weeks.

Clinical. Most infections are subclinical. Severe infections have been reported to cause retarded growth and weight loss in mice and intestinal occlusion, intestinal impaction, and death in hamsters.

Pathology. Presence of adult worms in the small intestine is usually associated with mild enteritis. Larval stages occasionally reach the lymph nodes, liver, or lung, where they incite a granulomatous inflammatory response.

Diagnosis. Diagnosis is made by demonstration and identification of adult tapeworms in the small intestine. Eggs can be demonstrated in feces. Also, histologic sections occasionally are successful in demonstrating the cercocystis in intestinal villi and lymph nodes.

Control. The most practical method of control is to obtain rodents from stocks demonstrated to be free of *H. nana*. Cesarean derivation and barrier maintenance are the most effective methods for eliminating infection.

Interference with Research. *H. nana* is a potential zoonotic infection to humans. It can interfere with studies involving the intestinal tract.

Suggested Reading

Flynn, R. J. 1973. Cestodes. Pp. 155-202 in Parasites of Laboratory Animals. Ames, Iowa: Iowa State University Press.

Hsu, C.-K. 1979. Parasitic diseases. Pp. 307-331 in The Laboratory Rat. Vol. I: Biology and Diseases, H. J. Baker, J. R. Lindsey, and S. H. Weisbroth, eds. New York: Academic Press.

Kunstyr, I., and K. T. Friedhoff. 1980. Parasitic and mycotic infections in laboratory animals. Pp. 181-192 in Animal Quality and Models in Biomedical Research, A. Spiegel, S. Erichsen, and H. A. Solleveld, eds. Stuttgart: Gustav Fischer Verlag.

Wescott, R. B. 1982. Helminths. Pp. 373-384 in The Mouse in Biomedical Research. Vol. II: Diseases, H. L. Foster, J. D. Small, and J. G. Fox, eds. New York: Academic Press.

Spironucleus muris

Agent. Flagellated protozoan, order Diplomonadida, family Hexamitidae, subfamily Hexamitinae. Formerly called *Hexamita muris*.

Life Cycle. Direct. Trophozoites reproduce by longitudinal fission and form highly resistant cysts. The minimal infectious dose for a mouse is one cyst.

Animals Affected. Mice, rats, and hamsters.

Epizootiology. Two- to 6-week-old mice are most susceptible to infection with *S. muris*. Trophozoites usually inhabit the crypts of Lieberkuhn in the small intestine, but in young animals the lumen can also contain large numbers of trophozoites. In older mice and in rats there are very few trophozoites, and those that are present can be found only in glands of the gastric pyloris. Transmission is by ingestion of cysts that are shed in the feces. The greatest numbers are shed by young or immunocompromised hosts. Cysts are inactivated by some disinfectants and high temperature (45°C for 30 minutes) but are highly resistant to most other environmental conditions. Infectivity is retained for 6 months at −20°C, for 1 day at pH 2.2, for 14 days at room temperature, or for 1 hour in 0.1% glutaraldehyde.

Clinical. Infection is usually subclinical in immunocompetent hosts. In athymic (*nu/nu*) and lethally irradiated mice, *S. muris* infection has been associated with severe chronic enteritis with weight loss.

Pathology. After ingestion of cysts, trophozoite (and cyst) numbers in the intestines of immunocompetent rodents peak at 1-2 weeks and decline to low numbers by 4-5 weeks in BALB/c mice; 7-9 weeks in CBA, SJL/J, and C3H/He mice; and 13 weeks in A and B.B10 mice. Numbers in athymic (*nu/nu*) mice persist indefinitely at high levels. In severe infections, the small intestine may appear reddened and contain watery fluid and gas. Smears of the intestinal contents contain numerous motile trophozoites, and cysts can be demonstrated in the cecum and colon. The best indicator of *S. muris* infection in hematoxylin and eosin-stained sections of the small intestine is distension of the crypts of Lieberkuhn by masses of granular-appearing trophozoites. Trophozoites can cause shortening of microvilli on the crypt epithelium and increased turnover of enterocytes. There is usually little or no inflammatory response in immunocompetent animals, but heavily parasitized, immunodeficient animals can have moderate to severe enteritis.

Diagnosis. Other possible causes of digestive tract disease (e.g., enterotrophic strains of mouse hepatitis virus) must be ruled out. Characteristic trophozoites can be demonstrated in the contents of the small intestine, or cysts can be demonstrated in the contents of the large intestine or feces using wet mounts under reduced light.

For routine health surveillance that includes histopathology, the examination of multiple histologic sections of the small intestine and gastric pylorus is probably superior to other methods because there may be very few parasites present, and they may be localized in distribution. Trophozoites can be stained by silver or periodic acid-Schiff methods.

Control. Cesarean derivation and barrier maintenance are recommended for control of the organism. Treatment of mice with 0.04-0.1% dimetridazole in drinking water for 14 days can ameliorate clinical signs but does not completely eliminate the infection.

Interference with Research. *S. muris* can increase the severity and mortality of the wasting syndrome (presumably due to mouse hepatitis virus) in athymic (*nu/nu*) mice. *S. muris* has been reported to increase mortality in cadmium-treated mice; alter macrophage function; reduce spleen plaque-forming cell responses to sheep erythrocytes; reduce lymphocyte responsiveness to mitogens such as phytohemagglutinin, concanavalin A, and pokeweed mitogen; and alter immune responsiveness to tetanus toxoid and type 3 pneumococcal polysaccharide. Whole-body irradiation increases susceptibility to *S. muris* infection and disease.

Suggested Reading

Brett, S. J. 1983. Immunodepression in *Giardia muris* and *Spironucleus muris* infections in mice. Parasitology 87:507-515.

Brett, S. J., and F. E. G. Cox. 1982. Interactions between the intestinal flagellates *Giardia muris* and *Spironucleus muris* and the blood parasites *Babesia microti*, *Plasmodium yoelii* and *Plasmodium berghei* in mice. Parasitology 85:101-110.

Keast, D., and F. C. Chesterman. 1972. Changes in macrophage metabolism in mice heavily infected with *Hexamita muris*. Lab. Anim. (London) 6:33-39.

Kunstyr, I., E. Ammerpohl, and B. Meyer. 1977. Experimental spironucleosis (hexamitiasis) in the nude mouse as a model for immunologic and pharmacologic studies. Lab. Anim. Sci. 27:782-788.

Stachan, R., and I. Kunstyr. 1983. Minimal infectious doses and prepatent periods in *Giardia muris*, *Spironucleus muris* and *Tritrichomonas muris*. Z. Bakt. Hyg. A256:249-256.

Wagner, J. E., R. E. Doyle, N. C. Ronald, R. G. Garrison, and J. A. Schmitz. 1974. Hexamitiasis in laboratory mice, hamsters, and rats. Lab. Anim. Sci. 24:349-354.

Syphacia obvelata (Mouse Pinworm) and *Syphacia muris* (Rat Pinworm)

Agents. Roundworms, order Ascarida, suborder Oxyurina.

Life Cycle. Direct; requires only 11-15 days for completion. Gravid females migrate from the large intestine to the perianal area, deposit their eggs, and then die. Eggs become infective in about 6 hours. Following ingestion by another host, eggs hatch in the small intestine, and the larvae reach the cecum in 24 hours. The parasites spend 10-11 days in the cecum where they mature and mate, thus continuing the cycle.

Animals Affected. Laboratory mice, rats, hamsters, gerbils, and wild rodents.

Epizootiology. Adults are found primarily in the cecum and colon of infected hosts. Eggs are efficiently disseminated from the perianal area of the host into the cage and room environments. The eggs can survive for weeks under most animal room conditions. Transmission is by ingestion of embryonated eggs.

Clinical. Infections caused by *Syphacia* spp. alone are subclinical.

Pathology. Pinworms of laboratory rodents are generally not considered pathogens. Pinworm burden in an infected rodent population is a function of age, sex, and host immune status. In enzootically infected colonies, weanling animals develop the greatest parasite loads, males are more heavily parasitized than females, and *Syphacia* numbers diminish with increasing age of the host. Athymic (*nu/nu*) mice have increased susceptibility to pinworm infection.

Diagnosis. Diagnosis is made by demonstrating eggs on the perianal region using the cellophane tape technique or by finding adult worms in the cecum and colon at necropsy.

Control. Cesarean derivation and barrier maintenance are effective methods of control. Hygienic methods, including frequent cage and room sanitization, can aid in controlling *Syphacia* in an infected rodent population. Cage-to-cage transmission can be prevented by using filter-top cages. Several anthelminthics are effective in eliminating a high percentage of adult worms but are inefficient in clearing immature worms or eggs.

Interference with Research. Pinworm infections in rats have been reported to reduce the occurrence of adjuvant-induced arthritis.

Suggested Reading

Flynn, R. J. 1973. Nematodes. Pp. 203-320 in Parasites of Laboratory Animals. Ames, Iowa: Iowa State University Press.

Hsu, C.-K. 1979. Parasitic diseases. Pp. 307-331 in The Laboratory Rat. Vol. I: Biology and Diseases, H. J. Baker, J. R. Lindsey, and S. H. Weisbroth, eds. New York: Academic Press.

Pearson, D. J., and G. Taylor. 1975. The influence of the nematode *Syphacia obvelata* on adjuvant arthritis in rats. Immunology 29:391-396.

Ross, C. R., J. E. Wagner, S. R. Wightman, and S. E. Dill. 1980. Experimental transmission of *Syphacia muris* among rats, mice, hamsters, and gerbils. Lab. Anim. Sci. 30:35-37.

Wescott, R. B. 1982. Helminths. Pp. 374-384 in The Mouse in Biomedical Research. Vol. II: Diseases, H. L. Foster, J. D. Small, and J. G. Fox, eds. New York: Academic Press.

Wescott, R. B., A. Malczewski, and G. L. Van Hoosier. 1976. The influence of filter top caging on the transmission of pinworm infections in mice. Lab. Anim. Sci. 26:742-745.

Trichomonas muris

Agent. Flagellated protozoan, order Trichomonadida.

Life Cycle. If a cyst stage exists, transmission is probably primarily by ingestion of cysts.

Animals Affected. Mice, rats, hamsters, and other rodents.

Epizootiology. Trophozoites are found throughout the fecal mass in the cecum and colon.

Clinical. Infections are subclinical.

Pathology. *T. muris* is considered a commensal.

Diagnosis. Diagnosis is by demonstration of trophozoites in wet mounts of contents from the cecum or colon. *T. muris* has characteristic wobbly or jerky movements. Trophozoites are found dispersed throughout the fecal stream in histologic sections of the cecum or colon prepared without disturbing the luminal contents. In hematoxylin and eosin-stained sections, the nucleus stains poorly, the nuclear membrane is indistinct, and the cell wall often appears wrinkled or folded upon itself.

Control. Control measures are usually not necessary.

Interference with Research. There have been no reports of interference with research results.

Suggested Reading

Hsu, C.-K. 1982. Protozoa. Pp. 359-372 in The Mouse in Biomedical Research. Vol. II: Diseases, H. L. Foster, J. D. Small, and J. G. Fox, eds. New York: Academic Press.

Kunstyr, I., B. Meyer, and E. Ammerpohl. 1977. Spironucleosis in nude mice: An animal model for immuno-parasitologic studies. Pp. 17-27 in Proceedings of the Second International Workshop on Nude Mice. Stuttgart: Gustav Fischer Verlag.

Levine, N. D. 1974. Diseases of laboratory animals—parasitic. Pp. 209-327 in CRC Handbook of Laboratory Animal Science, vol. II, E. C. Melby and N. H. Altman, eds. Cleveland: CRC Press.

Other Endoparasites

Numerous other endoparasites have been reported in wild mice and rats and are encountered occasionally in laboratory animals maintained by conventional methods. For information, comprehensive works on endoparasites should be consulted.

Suggested Reading

Flynn, R. J. 1973. Parasites of Laboratory Animals. Ames, Iowa: Iowa State University Press. 884 pp.

Griffiths, H. J. 1971. Some common parasites of small laboratory animals. Lab. Anim. (London) 5:123-135.

Hsu, C. K. 1979. Parasitic diseases. Pp. 307-331 in the Laboratory Rat. Vol. I: Biology and Diseases, H. J. Baker, J. R. Lindsey, and S. H. Weisbroth, eds. New York: Academic Press.

Hsu, C.-K. 1982. Protozoa. Pp. 359-372 in The Mouse in Biomedical Research. Vol. II: Diseases, H. L. Foster, J. D. Small, and J. G. Fox, eds. New York: Academic Press.

Levine, N. D. 1974. Diseases of laboratory animals—parasitic. Pp. 209-327 in CRC Handbook of Laboratory Animal Science, vol. II, E. C. Melby and N. H. Altman, eds. Cleveland: CRC Press.

Levine, N. D., and V. Ivens. 1965. The Coccidian Parasites (Protozoa, Sporozoa) of Rodents. Urbana, Ill.: University of Illinois Press. 365 pp.

Oldham, J. N. 1967. Helminths, ectoparasites and protozoa in rats and mice. Pp. 641-678 in Pathology of Laboratory Rats and Mice, E. Cotchin and F. J. C. Roe, eds. Oxford: Blackwell Scientific.

Wescott, R. B. 1982. Helminths. Pp. 374-384 in The Mouse in Biomedical Research. Vol. II: Diseases, H. L. Foster, J. D. Small, and J. G. Fox, eds. New York: Academic Press.

PART III

Diagnosis and ResearchComplications of Infectious Agents

INTRODUCTION

Part III is intended to serve as an index for diagnostic problem solving in situations for which infectious agents of mice and rats may be responsible. Three index categories, clinical signs, pathology, and research complications, are listed in alphabetical order. When a problem suspected of being caused by an infectious agent is encountered, one or more of these lists should be consulted to identify quickly the most likely causative agent(s). One should then consult the information in Part II on the most likely candidate agents to narrow the list of possible causes and to devise further testing to make the definitive diagnosis. Once the precise cause(s) is known, specific corrective measures can be implemented.

CLINICAL SIGNS

Abdominal Enlargement
    Kilham rat virus
    Leukemia viruses, murine
    Lymphocytic choriomeningitis virus

Abortions and Stillbirths
    Streptobacillus moniliformis

Abscesses
    Cervical
        Pasteurella pneumotropica
    Facial, orbital, and tail
        Staphylococcus aureus
    Preputial gland
        Pasteurella pneumotropica
        Staphylococcus aureus

Alopecia
    (See Dermatitis and Alopecia)

Amputations, Necrotic, of Limbs or Tails
    Corynebacterium kutscheri
    Ectromelia virus
    Mycoplasma arthritidis
    Ringtail
    Streptobacillus moniliformis

Annular Constrictions of Tail
    Ringtail

Anorexia
    Corynebacterium kutscheri
    Ectromelia virus

Ataxia
    Kilham rat virus

Athymic (*nu/nu*) Mice, More Susceptible than Immunocompetent Mice to:
    Cytomegalovirus, mouse
    Encephalitozoon cuniculi
    Giardia muris
    Hepatitis virus, mouse
    Pneumocystis carinii
    Pneumonia virus of mice
    Polyoma virus
    Sendai virus
    Spironucleus muris
    Staphylococcus aureus
    Syphacia spp.

Athymic (*nu/nu*) Mice, Equally or Less Susceptible than Immunocompetent Mice to:
    Bacillus piliformis
    Lymphocytic choriomeningitis virus
    Mycoplasma pulmonis
    Reovirus-3
    Streptococcus pneumoniae

Athymic (*nu/nu*) Mice, Not Susceptible to:
    Myobia musculi
    Myocoptes musculinus

Birth Weight Reduced
    Salmonella enteritidis
    Sendai virus

Cervical Edema
    Sialodacryoadenitis virus

Chattering (Mice)
    Mycoplasma pulmonis
    Sendai virus

Circling (or Rolling)
    Kilham rat virus
    Pseudomonas aeruginosa
    Streptobacillus moniliformis
    Theiler's virus

Conjunctivitis
    Ectromelia virus

Pasteurella pneumotropica
Salmonella enteritidis
Sialodacryoadenitis virus
Staphylococcus aureus
Streptobacillus moniliformis

Convulsions
    Theiler's virus

Corneal Ulceration
    Sialodacryoadenitis virus

Cyanosis
    Salmonella enteritidis
    Streptobacillus moniliformis

Deaths, High Mortality
(greater than 50%) Possible
    Bacillus piliformis
    Citrobacter freundii (Biotype 4280)
    Ectromelia virus
    Hepatitis virus, mouse (infant mice)
        Salmonella enteritidis
    Streptobacillus moniliformis
    Theiler's virus

Deaths in Neonates
    Hepatitis virus, mouse
    Sendai virus

DeathsUnlikely
(in uncomplicated infections)
    Adenoviruses, mouse
    Aspicularis tetraptera
    Coronavirus, rat
    Cytomegalovirus, mouse
    Encephalitozoon cuniculi
    Entamoeba muris
    Giardia muris
    Hantaviruses
    Hymenolepis nana
    H-1 virus
    Lactic dehydrogenase-elevating virus
    Minute virus of mice
    Mycoplasma arthritidis

Pasteurella pneumotropica
Pneumocystis carinii
Pneumonia virus of mice
Polyoma virus
Pseudomonas aeruginosa
Radfordia affinis
Sialodacryoadenitis virus
Spironucleus muris
Syphacia spp.
Theiler's virus
Thymic virus, mouse
Tritrichomonas muris

Deaths, Usually Low Mortality
    Corynebacterium kutscheri
    Hepatitis virus, mouse
    Kilham rat virus
    Lymphocytic choriomeningitis virus
    Mycoplasma pulmonis
    Rotavirus, mouse
    Salmonella enteritidis
    Sendai virus
    Streptococcus pneumoniae

Dehydration
    (See Diarrhea)

Dermatitis and Alopecia
    Due to Infectious Agents
        Dermatophytes (fungi)
        Ectromelia virus
        Myobia musculi
        Myocoptes musculinus
        Pasteurella pneumotropica
        Papule virus, mouse
        Staphylococcus aureus
    Due to Noninfectious Causes
        Bite (fight wounds)
        "Whisker trimming" ("hair nibbling," "barbering")

Diarrhea
    Bacillus piliformis
    Citrobacter freundii (Biotype 4280)

Giardia muris (evidence uncertain)
Hepatitis virus, mouse (infant mice)
Reovirus-3 (evidence uncertain)
Rotavirus, mouse (infant mice)
Salmonella enteritidis
Spironucleus muris (evidence uncertain)

Dyspnea
  Cilia-associated respiratory bacillus
  Corynebacterium kutscheri
  Leukemia virus, murine
  Mycoplasma pulmonis
  Sendai virus
  Streptococcus pneumoniae

Emaciation
  Lymphocytic choriomeningitis virus
  Salmonella enteritidis
  Streptobacillus moniliformis

Facial Abscesses
  Staphylococcus aureus

Facial Edema
  Ectromelia virus

Gestation Prolonged
  Sendai virus

Growth Retardation
  Cilia-associated respiratory bacillus
  Citrobacter freundii (Biotype 4280)
  Hymenolepis nana
  Kilham rat virus
  Lymphocytic choriomeningitis virus
  Mycoplasma pulmonis
  Reovirus-3 (evidence uncertain)
  Sendai virus
  Sialodacryoadenitis virus

Head Tilt
  Mycoplasma pulmonis
  Pseudomonas aeruginosa
  Theiler's virus

Hemoglobinuria
  Streptobacillus moniliformis

Hunched Posture
  (See Reluctance to Move)

Hyperexcitability
  Theiler's virus

Inapparent Infections (Agents that usually cause subclinical or latent infections under natural conditions)
  Adenoviruses, mouse
  Aspicularis tetraptera
  Bacillus piliformis
  Corynebacterium kutscheri
  Cytomegalovirus, mouse
  Dermatophytes
  Encephalitozoon cuniculi
  Entamoeba muris
  Giardia muris
  Hantaviruses
  Hepatitis virus, mouse
  Hymenolepis nana
  Kilham rat virus
  Lactic dehydrogenase-elevating virus
  Lymphocytic choriomeningitis virus
  Mammary tumor virus, mouse
  Minute virus of mice
  Rotavirus, mouse

Jaundice
  Hepatitis virus, mouse (athymic mice)
  Kilham rat virus
  Reovirus-3 (evidence uncertain)

Keratoconus
  Sialodacryoadenitis virus
  Kyphosis
  Streptobacillus moniliformis

Litter Size Reduced
  Kilham rat virus

Salmonella enteritidis
Sendai virus

Lymphadenopathy, Peripheral
   Leukemia viruses, mouse
   Myobia musculi
   Myocoptes musculinus

Mastitis
   Pasteurella pneumotropica

Ocular Discharge
   Sialodacryoadenitis virus

Pallor (Anemia)
   Streptobacillus moniliformis

Panophthalmitis
   Pasteurella pneumotropica

Papular Rash
   Ectromelia virus

Paralysis of Rear Legs
   Lactic dehydrogenase-elevating virus
      (in C58 and AKR mice)
   Polyomavirus [in athymic (*nu/nu*) mice]
   Streptobacillus moniliformis
   Theiler's virus

Photophobia
   Sialodacryoadenitis virus
   Streptobacillus moniliformis

Pododermatitis
   Staphylococcus aureus

Polypnea
   Cilia-associated respiratory bacillus
   Corynebacterium kutscheri
   Mycoplasma pulmonis
   Streptococcus pneumoniae

Priapism
   Streptobacillus moniliformis

Pruritis
   Dermatophytes (fungi)
   Myobia musculi
   Myocoptes musculinus
   Staphylococcus aureus

Rectal Prolapse
   Citrobacter freundii (Biotype 4280)
   Syphacia spp. (evidence uncertain)

Reluctance to Move (Animals often
sit in hunched posture and have
ruffled coats)
   Bacillus piliformis
   Ectromelia virus
   Lymphocytic choriomeningitis virus
   Mycoplasma pulmonis
   Salmonella enteritidis

Respiratory Rales
   Cilia-associated respiratory bacillus
   Corynebacterium kutscheri
   Mycoplasma pulmonis
   Streptococcus pneumoniae

Ruffled Hair Coat
   (See Reluctance to Move)

Runting
   (See Wasting Syndrome)

Scrotal Cyanosis
   Kilham rat virus

Self-Mutilation of Penis
   Staphylococcus aureus

Skin Ulceration
   Dermatophytes (fungi)
   Myobia musculi
   Myocoptes musculinus
   Staphylococcus aureus

Sneezing
   Klebsiella pneumoniae

Mycoplasma pulmonis
Sialodacryoadenitis virus

Snuffling
　Cilia-associated respiratory bacillus
　Mycoplasma pulmonis
　Sendai virus
　Sialodacryoadenitis virus
　Streptococcus pneumoniae

Stunted Growth
　(See Growth Retardation)

Subcutaneous Mass
　Mammary tumor virus, mouse

Swelling (Edema)
　Feet and tail
　　Ectromelia virus
　　Ringtail
　　Streptobacillus moniliformis
　Neck
　　Sialodacryoadenitis virus

Swollen, Reddened Joints
　Corynebacterium kutscheri
　Mycoplasma arthritidis
　Streptobacillus moniliformis

Tremors
　Theiler's virus

Wasting Syndrome
　Hepatitis virus, mouse [athymic (*nu/nu*) mice]
　Polyoma virus [athymic (*nu/nu*) mice]

Weight Loss
　Citrobacter freundii (Biotype 4280)
　Corynebacterium kutscheri
　Hymenolepis nana
　Kilham rat virus
　Mycoplasma pulmonis
　Salmonella enteritidis
　Sialodacryoadenitis virus
　Streptococcus pneumoniae

PATHOLOGY

Abscesses
    Cervical lymph nodes
        Pasteurella pneumotropica
    Face, orbits, and tail
        Staphylococcus aureus
    Kidney
        Corynebacterium kutscheri
    Liver
        Bacillus piliformis
        Corynebacterium kutscheri
    Lung
        Cilia-associated respiratory bacillus
        Mycoplasma pulmonis
    Preputial glands
        Pasteurella pneumotropica
        Staphlococcus aureus

Age-Dependent Polioencephalomyelitis
    Lactic dehydrogenase-elevating virus

Alopecia
    (See Dermatitis and Alopecia)

Amputations, Necrotic, of Limbs or Tails
    Corynebacterium kutscheri
    Ectromelia virus
    Mycoplasma arthritidis
    Ringtail
    Streptobacillus moniliformis
Amyloidosis
    Myobia musculi

Anemia
    Leukemia viruses, murine
    Streptobacillus moniliformis

Ankylosis
    Streptobacillus moniliformis

Arthritis
    Corynebacterium kutscheri

    Mycoplasma arthritidis
    Mycoplasma pulmonis [athymic
        (*nu/nu*) mice]
    Streptobacillus moniliformis

Ascites
    Encephalitozoon cuniculi
    Lymphocytic choriomeningitis virus

Atelectasis (Lung)
    Cilia-associated respiratory bacillus
    Mycoplasma pulmonis
    Sendai virus

Brain
    Cerebellar hypoplasia
        Kilham rat virus
    Choroiditis
        Lymphocytic choriomeningitis virus
    Ependymitis
        Lymphocytic choriomeningitis virus
    Glial nodules
        Encephalitozoon cuniculi
    Hemorrhage
        Kilham rat virus
    Intranuclear inclusions
        Kilham rat virus
    Leptomeningitis
        Encephalitozoon cuniculi
        Lymphocytic choriomeningitis virus
        Streptococcus pneumoniae
    Syncytial giant cells
        Hepatitis virus, mouse

Bronchiectasis and Bronchiolectasis
    Cilia-associated respiratory bacillus
    Mycoplasma pulmonis

Bronchiolar Scarring
    Cilia-associated respiratory bacillus
    Sendai virus

Bronchiolitis Obliterans
  Cilia-associated respiratory bacillus

Cell Fragment Thrombi
  Salmonella enteritidis

Cerebellar Hypoplasia
  Kilham rat virus

Colon and Cecum
  Cecocolitis
    Bacillus piliformis
  Colonic hyperplasia
    Citrobacter freundii (Biotype 4280)
  Goblet cell hyperplasia
    Citrobacter freundii (Biotype 4280)
  Helminth parasites
    Aspicularis tetraptera
    Syphacia spp.
  Hyperplastic typhlocolitis
    Hepatitis virus, mouse [athymic (*nu/nu*) mice]
  Protozoan parasites
    Entamoeba muris
    Tritrichomonas muris

Cutaneous Papules, Erosions or Encrustations
  Ectromelia virus

Demyelination and Remyelination
  Theiler's virus

Dermatitis and Alopecia
  (See Clinical Signs above)

Ear
  Otitis interna
    Mycoplasma pulmonis
    Pseudomonas aeruginosa
    Streptobacillus moniliformis
  Otitis media
    Cilia-associated respiratory bacillus
    Mycoplasma pulmonis
    Pasteurella pneumotropica
    Pseudomonas aeruginosa
    Streptococcus pneumoniae

Ectromelia (See Amputations, Necrotic, of Limbs or Tails)

Empyema
  Streptococcus pneumoniae

Encephalitis
  Granulomatous
    Encephalitozoon cuniculi
  Hemorrhagic
    Kilham rat virus
  Nonsuppurative
    Hepatitis virus, mouse
    (See also Polioencephalomyelitis)

Encephalomyelitis
  (See Polioencephalomyelitis)

Endometritis
  Mycoplasma pulmonis

Eosinophilic Crystals in Lung
  Sendai virus

Eye (conjunctivitis, corneal ulceration, hyphema, hypopyon, keratitis, keratoconus, lenticular degeneration, megaloglobus, pannus, retinal degeneration, synechia)
  Sialodacryoadenitis virus

Fetal Resorption
  Kilham rat virus
  Sendai virus
  Streptobacillus moniliformis

Glial Nodules, Brain
  Encephalitozoon cuniculi

Glomerulonephritis
  Embolic
    Corynebacterium kutscheri
  Immune complex
    Lactic dehydrogenase-elevating virus
    Lymphocytic choriomeningitis virus

Heart
  Myocarditis
    Bacillus piliformis
  Pericarditis
    Streptococcus pneumoniae

Hemoglobinuria
  Streptobacillus moniliformis

Hemorrhage
  Central nervous system,
  epididymis, and testes
    Kilham rat virus
  Jejunum
    Ectromelia virus
  Peyers patches
    Ectromelia virus

Hemorrhagic encephalopathy
  Kilham rat virus

Hepatic Necrosis
  Bacillus piliformis
  Corynebacterium kutscheri (mice)
  Hepatitis virus, mouse
  Kilham rat virus
  Lymphocytic choriomeningitis virus
  Mousepox virus
  Reovirus-3 (evidence uncertain)
  Salmonella enteritidis
  Streptobacillus moniliformis

Hepatitis, Bacterial
  Streptococcus pneumoniae

Hypersensitivity, Cutaneous
  Myobia musculi

  Myocoptes musculinus
  Staphylococcus aureus

Immune Complex Glomerulonephritis
  Lactic dehydrogenase-elevating virus
  Lymphocytic choriomeningitis virus

Inclusions
  Intracytoplasmic (skin)
    Ectromelia virus
  Intracytoplasmic and intranuclear
  (bronchi, ureters, and renal pelvis)
    Polyoma virus [athymic (*nu/nu*) mice]
  Intranuclear
    Harderian gland
      Sialodacryoadenitis virus
    Intestinal mucosa
      Adenovirus, mouse (MAd-2)
    Salivary gland
      Cytomegalovirus, mouse
    Thymus
      Thymic virus, mouse

Infarcts
  Central nervous system, epididymis, and testes
    Kilham rat virus
  Spleen and testes
    Streptococcus pneumoniae

Intestine, Small
  Blunting of villi
    Bacillus piliformis
    Hepatitis virus, mouse
    Rotavirus, mouse
  Enteritis
    Adenovirus, mouse (MAd-2)
    Bacillus piliformis
    Giardia muris
    Hepatitis virus, mouse
    Hymenolepis nana
    Reovirus-3 (evidence uncertain)
    Salmonella enteritidis

    Spironucleus muris
  Enteritis with ulceration
    Hepatitis virus, mouse
  Enteritis with ulceration and
  hemorrhage
    Bacillus piliformis
    Citrobacter freundii (Biotype 4280)
    Ectromelia virus
  Helminth parasite
    Hymenolepis nana
  Ileocecocolitis
    Bacillus piliformis
    Citrobacter freundii (Biotype 4280)
    Salmonella enteritidis
  Protozoan parasites
    Giardia muris
    Spironucleus muris
  Syncytial epithelial giant cells
    Hepatitis virus, mouse

Keratitis
  Sialoacryoadenitis virus

Kidney
  Cortical pitting and scarring
    Encephalitozoon cuniculi
  Nephritis
    Corynebacterium kutscheri
    Encephalitozoon cuniculi
  Protozoan parasite
    Encephalitozoon cuniculi

Lacrimal Glands (dacryoadenitis, intranuclear inclusions, necrosis, squamous metaplasia)
  Sialodacryoadenitis virus

Laryngitis
  Cilia-associated respiratory bacillus
  Mycoplasma pulmonis

Leukemia and Lymphomas
  Leukemia viruses, murine

Liver
  Biliary hyperplasia
    Kilham rat virus
  Intranuclear inclusions
    Kilham rat virus
  Inflammation, pyogranulomatous
    Salmonella enteritidis
  Necrosis
    Bacillus piliformis
    Corynebacterium kutscheri
    Ectromelia virus
    Hepatitis virus, mouse
    Kilham rat virus
    Lymphocytic choriomeningitis virus
    Salmonella enteritidis
    Streptobacillus moniliformis
    Streptococcus pneumoniae
  Thrombi
    Salmonella enteritidis

Lungs
  Abscesses
    Cilia-associated respiratory bacillus
    Corynebacterium kutscheri
    Mycoplasma pulmonis
  Atelectasis
    Cilia-associated respiratory bacillus
    Mycoplasma pulmonis
    Sendai virus
  Bronchiectasis and bronchiolectasis
    Cilia-associated respiratory bacillus
    Mycoplasma pulmonis
  Bronchiolitis, necrotizing, and granulomatous
    Cilia-associated respiratory bacillus
  Bronchiolitis obliterans
    Cilia-associated respiratory bacillus
  Bronchitis
    Cilia-associated respiratory bacillus
    Mycoplasma pulmonis
    Sendai virus
    Streptococcus pneumoniae
  Peribronchial and perivascular

lymphocyte cuffing
    Mycoplasma pulmonis
Pneumonia
    (see Pneumonia)
Squamoid change (peribronchiolar adematoid hyperplasia or alveolar bronchiolization)
    Sendai virus

Lymphadenopathy
    Generalized
        Leukemia virus, murine
        Streptobacillus moniliformis
    Mesenteric
        Bacillus piliformis
    Peripheral
        Leukemia virus, murine
        Myobia musculi
        Myocoptes musculinus

Lymph Nodes
    Abscesses
        Pasteurella pneumotropica
    Hyperplasia
        Bacillus piliformis
        Lymphocytic choriomeningitis virus
        Myobia musculi
        Myocoptes musculinus
        Salmonella enteritidis
    Necrosis
        Ectromelia virus

Lymphocytopenia
    Lactic dehydrogenase-elevating virus

Mammary Adenocarcinoma and Carcinosarcoma
    Mammary tumor virus, mouse

Mastitis
    Pasteurella pneumotropica
Mediastinitis
    Streptococcus pneumoniae

Megaloileitis in Rats
    Bacillus piliformis

Meningitis, Suppurative
    Streptococcus pneumoniae

Meningoencephalitis
    Encephalitozoon cuniculi
    Lymphocytic choriomeningitis virus

Necrotizing Tracheitis, Bronchitis, and Bronchiolitis
    Sendai virus

Neoplasia
    Leukemias and lymphomas
        Leukemia viruses, murine
    Mammary glands
        Mammary tumor virus, mouse
    Salviary glands (also kidney, subcutis, mammary gland, adrenal, bone, cartiliage, blood vessels, and thyroid)
        Polyoma virus (experimental)

Orchitis, Suppurative
    Streptococcus pneumoniae

Ovary, Perioophoritis, Purulent
    Mycoplasma pulmonis

Peribronchiolar Adenomatous Hyperplasia (synonym: adenomatous change, alveolar bronchiolization)
    Sendai virus

Pericarditis
    Streptococcus pneumoniae

Perioophoritis
    Mycoplasma pulmonis

Peritonitis
    Corynebacterium kutcheri
    Salmonella enteritidis
    Streptococcus pneumoniae

Pleuritis
   Corynebacterium kutscheri
   Streptococcus pneumoniae

Pleural Effusion, Serous
   Sendai virus

Pneumonia
   Bronchopneumonia
      Cilia-associated respiratory bacillus
      Mycoplasma pulmonis
      Streptococcus pneumoniae
   Embolic
      Corynebacterium kutscheri
   Interstitial
      Hepatitis virus, mouse
      Pneumocystis carinii
      Pneumonia virus of mice, (experimental)
      Sendai virus
      Sialodacryoadenitis virus
   Lobar
      Streptococcus pneumoniae
   Necropurulent
      Corynebacterium kutscheri
   Unspecified type
      Streptococcus pyogenes

Pododermatitis, Traumatic
   Staphylococcus aureus

Polioencephalomyelitis
   Lactic dehydrogenase-elevating virus (C58 and AKR mice)
   Polyoma virus [athymic (*nu/nu*) mice]
   Theiler's virus

Polyarthritis
   Mycoplasma arthritidis

Pox
   Ectromelia virus

Preputial Gland Abscesses
   Staphylococcus aureus
   Pasteurella pneumontropica

Pseudotuberculosis
   Corynebacterium kutscheri

Pyometra
   Mycoplasma pulmonis

Rhinitis
   Cilia-associated respiratory bacillus
   Corynebacterium kutscheri
   Mycoplasma pulmonis
   Sendai virus
   Sialodacryoadenitis virus
   Streptococcus pneumoniae

Salivary Glands
   Intranuclear inclusions
      Cytomegalovirus, mouse
   Necrosis
      Sialodacryoadenitis virus
   Neoplasms
      Polyomavirus
   Sialodenitis
      Sialodacryoadenitis virus

Salpingitis
   Mycoplasma pulmonis

Scrotal Hemorrhage
   Kilham rat virus

Septicemia
   Bacillus piliformis
   Corynebacterium kutscheri
   Salmonella enteritidis
   Streptobacillus moniliformis
   Streptococcus pneumoniae
Skin
   Ectoparasites
      Myobia musculi

Myocoptes musculinus
Radfordia affinis
Papules
  Ectromelia virus
Pox
  Ectromelia virus

Spinal Cord Hemorrhage
  Kilham rat virus

Spleen
  Infarction
    Streptococcus pneumoniae
  Necrosis, Diffuse
    Ectromelia virus
  Necrosis, Multifocal
    Ectromelia virus
    Salmonella enteritidis
    Streptobacillus moniliformis
  Thrombi
    Salmonella enteritidis
    Streptococcus pneumoniae

Splenitis
  Acute bacterial
    Streptococcus pneumoniae
  Pyogranulomatous
    Salmonella enteritidis

Splenomegaly
  Corynebacterium kutscheri
  Ectromelia virus
  Hepatitis virus, mouse
  Lymphocytic choriomeningitis virus
  Myobia musculi
  Myocoptes musculinus
  Salmonella enteritidis
  Streptobacillus moniliformis
  Streptococcus pneumoniae

Synctial Giant Cells
  Brain
    Hepatitis virus, mouse

Bronchial epithelium
  Mycoplasma pulmonis (mouse)
  Sendai virus
Intestinal epithelium
  Hepatitis virus, mouse
Multiple organs
  Hepatitis virus, mouse [athymic (*nu/nu*) mice]
Nasal epithelium
  Mycoplasma pulmonis (mouse)

Teratogenic Effects
  Kilham rat virus

Testes
  Hemorrhage
    Kilham rat virus
  Infarction
    Kilham rat virus
    Streptococcus pneumoniae

Thrombocytopenia
  Cytomegalovirus, mouse

Thrombosis
  Central nervous system, epididymis, and testes
    Kilham rat virus
  Liver and spleen
    Salmonella enteritidis
  Spleen and testes
    Streptococcus pneumoniae

Thymus
  Enlargement
    Leukemia viruses, murine
  Necrosis
    Ectromelia virus
    Lactic dehydrogenase-elevating virus
    Sialodacryoadenitis virus
    Thymic virus, mouse

Tracheitis
  Cilia-associated respiratory bacillus

Corynebacterium kutscheri
Mycoplasma pulmonis
Sendai virus
Sialodacryoadenitis virus
Streptococcus pneumoniae

Typhlocolitis
   Hepatitis virus, mouse [athymic (*nu/nu*) mice]

Ulcerative Cecitis in Rats
   Salmonella enteritidis

Ulcerative Dermatitis
   Myobia musculi

   Myocoptes musculinus
   Staphylococcus aureus

Uterus
   Endometritis
      Mycoplasma pulmonis
   Fetal resorption
      Kilham rat virus
      Sendai virus
      Streptobacillus piliformis
   Pyometra
      Mycoplasma pulmonis

RESEARCH COMPLICATIONS

Altered Immune Response
   Cytomegalovirus, mouse
   Ectromelia virus
   Encephalitozoon cuniculi
   Giardia muris
   Hepatitis virus, mouse
   Kilham rat virus
   Lactic dehydrogenase-elevating virus
   Lymphocytic choriomeningitis virus
   Minute virus of mice
   Mycoplasma pulmonis
   Myobia musculi
   Myocoptes musculinus
   Salmonella enteritidis
   Sendai virus
   Spironucleus muris
   Syphacia spp.
   Thymic virus, mouse

Altered Physiologic, Pharmacologic,
or Toxicologic Response
   Bacillus piliformis
   Hepatitis virus, mouse
   Kilham rat virus
   Lactic dehydrogenase-elevating virus
   Mycoplasma pulmonis
   Salmonella enteritidis
   Streptococcus pneumoniae

Altered Susceptibility to Other
Infections
   Cytomegalovirus, mouse
   Encephalitozoon cuniculi
   Hepatitis virus, mouse
   Lactic dehydrogenase-elevating virus
   Lymphocytic choriomeningitis virus
   Mycoplasma pulmonis
   Rotavirus, mouse
   Salmonella enteritidis
   Sendai virus
   Sialodacryoadenitis virus

Carcinogenesis or Spontaneous
Neoplasia
   Citrobacter freundii (Biotype 4280)
   H-1 virus
   Lactic dehydrogenase-elevating virus
   Lymphocytic choriomeningitis virus
   Mammary tumor virus, mouse
   Mycoplasma pulmonis
   Polyoma virus
   Sendai virus

Contamination of Cell Cultures
   Kilham rat virus
   Lymphocytic choriomeningitis virus
   Minute virus of mice
   Mycoplasma arthritidis
   Mycoplasma pulmonis
   Polyoma virus
   Reovirus-3

Contamination of Transplantable
Tumors and Altered Host Response
   Encephalitozoon cuniculi
   H-1 virus
   Kilham rat virus
   Lactic dehydrogenase-elevating virus
   Lymphocytic choriomeningitis virus
   Minute virus of mice
   Mycoplasma arthritidis
   Mycoplasma pulmonis
   Polyoma virus
   Reovirus-3
   Sendai virus

Inapparent Infection Exacerbated by
Experimental Immunosuppression
   Bacillus piliformis
   Corynebacterium kutscheri
   Ectromelia virus
   Giardia muris
   Hepatitis virus, mouse

Kilham rat virus
Mycoplasma pulmonis
Pneumocystis carinii
Pseudomonas aeruginosa
Salmonella enteritidis
Spironucleus muris

Zoonotic Agents (Infectious for Humans)
    Dermatophytes (fungi)
    Hantaviruses
    Hymenolepis nana
    Lymphocytic choriomeningitis virus
    Salmonella enteritidis
    Streptobacillus moniliformis
    Streptococcus pneumoniae

Index

A

Abdomen
  abdominal breathing, 54
  enlargement, 28, 44, 71
  lesions, 54
Abortions and stillbirths, 53, 71
Abscesses, 12, 14, 17, 25, 38, 52, 53, 71, 73, 76
Adenoviruses, 9–10
Age factors
  bacterial infections, 10, 13
  viral infections, 22, 30, 44, 48, 76
Agents (characterization), *see* Bacterial infections: agent characterization; Viral agents; Zoonotic agents
Airflow systems, 41
Alopecia, *see* Dermatitis and alopecia
Amputations, 71, 76
Amyloidosis, 59, 76
Anemia, 23, 53, 76
Ankylosis, 53, 76
Anorexia, 15, 71
Arthritis, 35, 37, 53, 54, 76
Ascites, 30, 76

Aspicularis tetraptera, 61
Ataxia, 71
Atelectasis, 12, 48, 76
Athymic (*nu/nu*) mice, immune response, 10, 15, 22, 23, 30–31, 36, 39, 40, 41, 48, 52, 62, 63, 65, 66, 71

B

Bacillus piliformis, 10–12
Bacterial infections
  age factors, 10, 13
  agent characterization, 1, 4, 10, 12, 13, 38, 42, 45, 51, 53, 54
  arthritis, 35, 37, 53, 54, 76
  barrier systems, 11, 13, 14, 36, 53, 54
  bronchiectasis and bronchiolectasis, 12, 36
  carcinogenesis/spontaneous neoplasia, 11, 37
  cell/tissue cultures, 35, 37, 46, 52
  cesarean control, 11, 12, 13, 14, 15, 36, 37, 38, 43, 47, 53, 54
  clinical signs, 10, 12, 13, 14–15, 35, 36, 38, 41, 42, 45–46, 52, 54
  colon and cecum, 10, 11, 14, 46

control techniques, general, 13, 37, 38, 47, 52, 53, 54
diagnosis, general, 11, 13, 14, 15, 35, 36–37, 38, 41–42, 46, 51
digestive system, 42–43, 45–46, 52
ectoparasites, secondary infection, 59
ELISA, 13, 15, 35, 37, 38
epizootiology, 10, 13, 14, 36, 38, 42, 45, 52, 53, 54
eyes, 11
hyperplasia, 14, 36
immune response, 10, 14, 36, 42, 43
intestines, 10, 13, 38, 47
liver, 10, 11, 15, 46, 47, 53
lymph nodes, 11, 14, 46
mortality, 10, 13, 37, 42, 43, 53
pathology, general, 10, 13–14, 15, 36, 38, 42, 46, 52, 53, 54
reluctance to move, 11, 14, 45, 54
research complications, general, 11, 13, 14, 15, 35, 36, 43, 47, 52, 53, 54
respiratory, 12, 14, 15, 36–37, 38, 53, 54
skin and joints, 38, 52, 53
stain techniques, 11, 12, 13
subclinical, 10, 11, 14, 15, 35, 36, 37, 38, 42, 53, 54
tumors/transplants, 35, 37
weight loss, 13, 14, 36, 45–46, 53, 54
X irradiation, 42, 43
zoonotic, 42, 45, 47, 51, 53, 54
see also specific agents
Barrier techniques
bacterial infections, 11, 13, 14, 36, 53, 54
fungi, 57, 58
parasites, 59, 63, 64
viral infections, 17, 22, 31, 33, 40, 44, 51, 55
Behavioral trauma, 59
chattering (mice), 36, 48, 71

circling and rolling, 42, 55, 71
head tilt, 11, 36, 73
hyperexcitability, 55, 73
reluctance to move, 11, 14, 30, 45, 54, 74
self-mutilation, 52, 74
Brain, 19, 22, 42, 76, 77
Breeding, 2, 23, 49, 51
see also Cesarean derivation
Bronchiectasis and bronchiolectasis, 76
bacterial infections, 12, 36
viral infections, 48

C

Carcinogenesis/spontaneous neoplasia, 84
bacterial infections, 11, 37
leukemia, 28–29, 79
mammary glands, 28, 32–33
Cell/tissue cultures, 4, 84
bacteria, 35, 37, 46, 52
viruses, 9, 51
Cervical edema, 50, 71
Cesarean derivation, 2
bacterial infections, 11, 12, 13, 14, 15, 36, 37, 38, 43, 47, 53, 54
fungi, 57
parasites, 59, 63, 64, 67
viral infections, 9, 18, 22, 25–26, 34, 40, 41, 44, 49, 56
Chattering (mice), 36, 48, 71
Cilia-associated respiratory bacillus (CAR), 12–13, 36
Circling and rolling, 42, 55, 71
Citrobacter freundii, 13–14
Clinical signs, 71–75
bacterial infections, 10, 12, 13, 14–15, 35, 36, 38, 41, 42, 45–46, 52, 54
fungi, 57–58
parasites, 59, 62, 64, 65, 67
viral infections, 17, 24–33, 40, 41, 44, 48, 50, 55
Colon and cecum, 77

bacterial infections, 10, 11, 14, 46
parasites, 66, 67, 68
viral infections, 44, 48
Complement fixation (CF), 3
bacterial infections, 11
viral infections, 44, 55
Conjunctivitis, 38, 46, 50, 53, 71–72
Control techniques
airflow systems, 41
bacterial infections, 13, 37, 38, 47, 52, 53, 54
fungi, 57–58
gnotobiotic animal and methods, 39, 43
parasites, 59, 63, 64, 66, 67, 68
pathogen-free animals, 2, 27
protozoa, 19, 39, 66, 67, 68
vaccination, 17, 49
viral infections, 16, 17–18, 21, 22, 24, 25–26, 27, 31, 33, 34, 41, 48, 49, 51, 56
see also Barrier techniques; Cesarean derivation; Specific pathogen-free animals
Convulsions, 55, 72
Corneal ulceration, 72
Corynebacterium kutscheri, 14–15, 53
Cyanosis, 53, 72
Cytomegalovirus, 15–16

D

Dacryoadenitis, 38
Deaths, high and low mortality, 72
bacterial infections, 10, 13, 37, 42, 43, 53
parasites, 64
viral infections, 17, 30, 55
Dehydration, see Diarrhea
Demyelination and remyelination, 77
Dermatitis and alopecia, 52, 59, 71, 72, 76
pododermatitis, 74, 81
Dermatophytes, 57–58

Diabetes, 31
Diagnosis, 3, 4, 70–83
bacterial infections, 11, 13, 14, 15, 35, 36–37, 38, 41–42, 46, 51
DNA probes, 13
fungi, 57–58
laboratories for, 6–7
parasites, 59, 62, 63, 64, 65, 67, 68
protozoa, 19, 39, 65–66, 67, 68
stain techniques, 11, 12, 13, 39
viral infections, 9, 16, 17, 21, 22, 25, 27, 29, 31, 33, 34, 40, 41, 43–44, 44–45, 48–49, 51, 56, 57
see also Cell/tissue cultures; Clinical signs; Pathology; Serologic tests
Diarrhea, 44, 46, 53, 72, 73
Digestive system infections
bacterial agents, 42–43, 45–46, 52
parasites, 62, 63, 64, 65, 66, 67
viral agents, 22
see also Intestines; *specific agents and organs*
DNA probes, 13
DNA viruses (agent characterization), 9, 15, 16, 24, 25, 33, 41
see also specific agents
Dyspnea, 14, 54, 73

E

Ear pathologies, 77
otitis media, 36, 38, 42, 77
Ectoparasites, 58–60
Ectromelia virus, 16–18, 53
Empyema, 54, 77
Encephalitis, 19, 77
Encephalitozoon cuniculi, 18–20
Encephalomyelitis, 28, 55–56, 77, 81
Endoparasites, 61–69
Entamoeba muris, 61–62
Enzyme-linked immunosorbent assay (ELISA), 3–4
bacterial infections, 13, 15, 35, 37, 38

viral infections, 9, 16, 17, 21, 22, 24, 25, 31, 34, 40, 41, 43, 44, 48–49, 51, 56, 57
Eosinophilic crystals in lung, 77
Epizootiology
   bacterial infections, 10, 13, 14, 36, 38, 42, 45, 52, 53, 54
   fungi, 57
   parasites, 59, 61, 62, 64, 65, 67, 68
   protozoa, 19, 39, 65, 67, 68
   viral infections, 9, 15–17, 20, 22, 24, 25, 26, 28, 30, 32, 33, 40, 41, 43, 44, 46, 50, 55, 57
Escherichia coli, 9
Eyes, 77
   bacterial infections, 11
   conjunctivitis, 38, 46, 50, 53, 71–72
   lacrimal glands, 50, 51, 79

F

Fungi, 39–40, 57–58, 59
Fur, *see* Hair

G

Germfree animal, *see* Pathogen-free animal
Giardia muris, 62–64
Gnotobiotic animal and methods, 39, 43
Growth retardation, 64, 73

H

H-1 virus, 24–25
Hair, 52, 57, 58
   mites, 58–60
   ruffled, 17, 36, 45, 74
Hantaviruses, 20–21
Head tilt, 11, 36, 73
Heart, 10, 78
Hemagglutination inhibition test (HAI), 3, 4, 17, 21, 24, 25, 41, 56
Hemoglobinuria, 53, 73
Hemorrhages, 20, 25, 78
Hepatic system, *see* Liver

Hepatitis, 54, 78
Hepatitis virus, 21–24, 63, 66
Herpesviruses, 56–57
Hosts
   bacterial infection, 52
   viral infection, 20
   see also Transplants, tissues and organs; Zoonotic agents
Hunched posture, *see* Reluctance to move
Hymenolepis nana, 64–65
Hyperexcitability, 55, 73
Hyperplasia
   bacterial infections, 14, 36
   ectoparasites, 59
   viral infections, 30
Hypersensitivity, cutaneous, 78

I

Immune response, 39, 84–85
   athymic (*nu/nu*) mice, 10, 15, 22, 23, 30–31, 36, 39, 40, 41, 48, 52, 62, 63, 65, 66, 71
   bacterial infections, 10, 14, 36, 42, 43
   parasites, 62, 63, 65
   viral infections, 15, 17, 22, 23, 26, 27, 30, 31, 33, 40, 41, 49
   see also Serologic tests
Inapparent infections, *see* Subclinical infections
Inclusions, 78
Indirect immunofluorescent antibody test (IFA), 3
   bacterial infections, 11, 12
   protozoa, 19, 39
   viral infections, 9, 16, 21, 22, 25, 31, 34, 44, 51, 57
Infarcts, 78
Intestines, 78–79
   bacterial infections, 10, 13, 38, 47
   parasites, 62, 63, 64, 65, 66
   viral infections, 9, 22, 44, 55

J

Jaundice, 73
Joints, *see* Skin and joints

K

Keratitis, 79
Keratoconus, 50, 73
Kidneys, 9, 15, 20, 43, 44, 79
Kilham rat virus, 25–26
Kyphosis, 53, 73

L

Lacrimal glands, 50, 51, 79
Lactic dehydrogenase-elevating virus, 26–28
Laryngitis, 36, 79
Leukemia, 28–29, 79
Life cycles, parasites, 58–59, 61, 62, 64, 65, 66, 67
Litter size reduced, 74
Liver, 79
    bacterial infections, 10, 11, 15, 46, 47, 53
    hepatitis, 21–24, 54
    jaundice, 73
    parasites, 64
    viral infections, 9, 17, 21–24, 25, 30, 54, 63, 66
Lung pathology, 39, 77, 79–80
    bacterial infections, 12, 15
    eosinophilic crystals, 77
    parasites, 64
    pneumonia, 5, 38–41, 48, 54–55, 81
    viral infection, 20, 21, 48, 51
Lymph nodes, 80
    bacterial infections, 11, 14, 46
    parasites, 59, 64
    viral infections, 17, 50–51, 55, 57
Lymphadenopathy, 53, 74, 80
Lymphocytic choriomeningitis virus, 30–32
Lymphocytopenia, 80

M

Mammary tumors, 28, 32–33, 80
Mastitis, 38, 74, 80
Microscopic procedures, 4, 59
Microsporum spp., 57–58
Minute virus of mice, 33–34
Mites, 58–60
Monitoring and monitored animals, 2, 3–6
Mouse hepatitis virus, 21–24, 63, 66
Mouse mammary tumor virus, 32–33
Mouse minute virus, 33–34
Mouse rotavirus, 44–45
Mouse thymic virus, 56–57
Murine respiratory mycoplasmosis (MRM), 36–38
Mycoplasma arthritidis, 35
Mycoplasma musculinus, 60
Mycoplasma pulmonis, 36–38
Myobia musculus, 60
Myocoptes musculinus, 59

N

Neoplasia, 27–28, 80
    leukemia, 28–29, 79
    mammary glands, 28, 32–33
    see also Carcinogenesis/spontaneous neoplasia

O

Otitis media, 36, 38, 42, 77

P

Pallor, *see* Anemia
Panophthalmitis, 38, 74
Paralysis, 42, 53, 74
Parasitic infections, 19, 55
    barrier techniques, 59, 63, 64
    cesarean derivation, 59, 63, 64, 67
    clinical signs, general, 59, 62, 64, 65, 67
    colon and cecum, 66, 67, 68
    control techniques, 59, 63, 64, 66, 67, 68

diagnosis, 59, 62, 63, 64, 65, 67, 68
digestive system, 62, 63, 64, 65, 66, 67
ectoparasites, 58–60
endoparasites, 61–69
epizootiology, 59, 61, 62, 64, 65, 67, 68
immune response, 62, 63, 65
intestinal, 62, 63, 64, 65, 66
lymph nodes, 59, 64
pathology, general, 59, 60, 62–63, 64, 65, 67, 68
research complications, 59, 60, 63, 64, 66, 67, 68
respiratory system, 36, 64
skin and joints, 59–60
subclinical infections, 62, 64, 67
weight loss, 62, 64
X irradiation, 65, 66
zoonotic, 62, 64
see also Life cycles, parasites; Protozoa
Pasteurella pneumotropica, 38–39
Pathogen-free animals, 2, 27
gnotobiotic animal and methods, 39, 43
see also Specific pathogen-free animals
Pathology, 76–83
bacterial infections, 10, 13–14, 15, 36, 38, 42, 46, 52, 53, 54
fungi, 57–58
parasites, 59, 60, 62–63, 64, 65, 67, 68
protozoa, 19, 39, 65, 67, 68
viral infections, 9, 16, 17, 21, 22, 24, 25, 27, 29, 30–31, 32–33, 40, 41, 44, 48, 50–51, 55
see also Clinical signs; *specific organs and systems*
Peritonitis, 46, 80–81
Photophobia, 50, 53, 74
Pinworm, 61, 66–67

Pleuritis, 54, 81
Pneumocystis carinii, 39–40
Pneumonia, 5, 38–41, 48, 54–55, 81
Pneumonia virus of mice, 5, 40–41
Pododermatitis, 74, 81
Polioencephalomyelitis, *see* Encephalomyelitis
Polyoma virus, 41–42
Polypnea, 36, 74
Poxviruses, 81
Preputial gland abscesses, 52, 81
Priapism, 53, 74
Protozoa, 18–20, 39–40, 65, 67–68
Pruritis, 59, 74
Pseudomonas aeruginosa, 42–43
Pseudotuberculosis, 81

R

Radfordia affinis, 60
Radioimmunoassays, 44
Rectal prolapse, 13, 74
Reluctance to move, 74
bacterial infections, 11, 14, 45, 54
viral infections, 30
Reovirus-3, 43–44
Research complications, 70, 84–85
bacterial infections, 11, 13, 14, 15, 35, 36, 43, 47, 52, 53, 54
parasites, 59, 60, 63, 64, 66, 67, 68
protozoa, 19, 39, 66, 67, 68
viral infections, 9, 18, 23, 24, 26, 27–28, 29, 31, 33, 34, 41–42, 44, 49, 51, 56, 57
see also Zoonotic agents
Respiratory rales, 14, 36, 74
Respiratory system, 39–40
bacterial infections, 12, 14, 15, 36–37, 38, 53, 54
bronchiectasis and bronchiolectasis, 12, 36, 48, 76
dyspnea, 14, 54, 73
parasites, 36, 64
sneezing, 50, 74

snuffling, 36, 54, 74
tracheitis, 83
viral infections, 21, 22, 47–48, 51, 55
see also Lungs
Rhinitis, 36, 54, 81
RNA viruses (agent characterization), 20, 21, 26, 28, 30, 32, 40, 43, 44, 47, 50, 55
see also specific agents
Rolling, *see* Circling and rolling
Rotavirus, 44–45
Runting, *see* Wasting syndrome

S

Salivary glands and saliva, 81
  viral infections, 15, 16, 28, 41, 50, 51, 56, 57
Salmonella enteritidis, 5, 45–47, 53
Salpingitis, 81
Sampling, 4–6
Scrotal cyanosis, 74
Scrotal hemorrhage, 81
Self-mutilation, 52, 74
Sendai virus, 5, 36, 47–50
Sentinel animals, 6, 23
Septicemia, 15, 42, 43, 53, 54, 81
Serologic tests, 3–4, 19, 46, 56, 59
  hemagglutination inhibition test (HAI), 3, 4, 17, 21, 24, 25, 41, 56
  indirect immunofluorescent antibody test (IFA), 3, 9, 11, 12, 16, 21, 19, 39, 22, 25, 31, 34, 44, 51, 57
  see also Enzyme-linked immunosorbent assay
Sialodacryoadenitis virus, 36, 50–51
Silver stain, *see* Stain techniques
Skin and joints, 75, 82
  arthritis, 35, 37, 53, 54, 76
  bacterial agents, 38, 52, 53
  fungi, 57–58
  hypersensitivity, 78

  parasites, 59–60
  viral agents, 15, 16–18
  see also Dermatitis and alopecia
Sneezing, 50, 74
Snuffling, 36, 54, 74
Social behavior, *see* Behavioral trauma
Specific pathogen-free animals, 2, 47
  viruses, 21, 25, 27, 33, 34, 49, 51, 56
Spinal cord, 82
Spironucleus muris, 65–66
Spleen, 17, 46, 53, 54, 57, 59, 66, 82
Stain techniques, 39
  bacterial infections, 11, 12, 13
Standards, 2, 3
Staphylococcus aureus, 51–53, 59
Streptobacillus moniliformis, 53–54
Streptococcus pneumoniae, 54–55
Subclinical infections, 4, 5, 39
  bacterial, 10, 11, 14, 15, 35, 36, 37, 38, 42, 53, 54
  fungi, 57
  parasites, 62, 64, 67
  viral, 16, 20, 23, 24, 30, 34, 40, 43, 55, 56
Symptoms, *see* Clinical signs; *specific symptoms*
Syphacia obvelata/muris, 66–67
Synctal giant cells, 82

T

Teratogenic effects, 82
Testes, 54, 82
Theiler's virus, 55–56
Thrombosis, 82
Thymic virus, 56–57
Thymus, 50–51, 56–57, 82–83
  athymic (*nu/nu*) mice, immune response, 10, 15, 22, 23, 30–31, 36, 39, 40, 41, 48, 52, 62, 63, 65, 66, 71
  lymphomas, 28, 29
Thyroid gland, 54

Tissue cultures, *see* Cell/tissue cultures
Tracheitis, 83
Transplants, tissues and organs, 84
  bacteria, 35, 37
  protozoa, 19
  viral infections, 17, 18, 21, 27, 31, 34, 41, 44
Tremors, 55
Trichomonas muris, 67–68
Trichophyton spp., 57
Tumors, 84
  bacteria, 35, 37
  protozoa, 19
  viral infections, 17, 18, 20, 21, 27, 31, 32, 34, 41, 42, 43
  see also Neoplasia
Tyzzer's disease, 10, 11, 46

U

Uterine pathologies, 38, 48, 83

V

Vaccination, 17, 49
Viral agents, 2, 9–10, 15, 56
  age factors, 22, 30, 44, 48, 76
  barrier techniques, 17, 22, 31, 33, 40, 44, 51, 55
  bronchiectasis and bronchiolectasis, 48
  cesarean derivation, 9, 18, 22, 25–26, 34, 40, 41, 44, 49, 56
  clinical signs, general, 17, 24–33, 40, 41, 44, 48, 50, 55
  colon and cecum, 44, 48
  complement fixation, 44, 55
  control techniques, general, 16, 17–18, 21, 22, 24, 25–26, 27, 31, 33, 34, 41, 48, 49, 51, 56
  culture techniques, 9, 51
  diagnosis, 9, 16, 17, 21, 22, 25, 27, 29, 31, 33, 34, 40, 41, 43–44, 44–45, 48–49, 51, 56, 57

digestive system, 9, 22, 44, 55
ELISA, 9, 16, 17, 21, 22, 24, 25, 31, 34, 40, 41, 43, 44, 48–49, 51, 56, 57
epizootiology, 9, 15–17, 20, 22, 24, 25, 26, 28, 30, 32, 33, 40, 41, 43, 44, 46, 50, 55, 57
HAI test, 17, 21, 24, 25, 41, 56
IFA test, 9, 16, 21, 22, 25, 31, 34, 44, 51, 57
immune response, 15, 17, 22, 23, 26, 27, 30, 31, 33, 40, 41, 49
liver, 9, 17, 21–24, 25, 30, 54, 63, 66
lung, 20, 21, 48, 51
lymph nodes, 17, 50–51, 55, 57
mortality, 17, 30, 55
pathology, general, 9, 16, 17, 21, 22, 24, 25, 27, 29, 30–31, 32–33, 40, 41, 44, 48, 50–51, 55
research complications, 9, 18, 23, 24, 26, 27–28, 29, 31, 33, 34, 41–42, 44, 49, 51, 56, 57
respiratory system, 21, 22, 47–48, 51, 55
salivary glands, 15, 16, 28, 41, 50, 51, 56, 57
skin and joints, 15, 16–18
specific pathogen-free animals, 21, 25, 27, 33, 34, 49, 51, 56
subclinical infections, 16, 20, 23, 24, 30, 34, 40, 43, 55, 56
transplants, 17, 18, 21, 27, 31, 34, 41, 44
tumors, 17, 18, 20, 21, 27, 31, 32, 34, 41, 42, 43
uterine, 48
X irradiation, 17
zoonotic, 20, 21, 30, 31
see also DNA viruses (agent characterization); RNA viruses (agent characterization); *specific agents*

W

Wasting syndrome, 30, 40, 42, 48, 53, 63, 73, 75
   anorexia, 15, 71
   see also Weight loss
Weight loss, 39, 71, 73, 75
   bacterial infections, 13, 14, 36, 45–46, 53, 54
   parasites, 62, 64

X

X-irradiation
   bacterial infections, 42, 43
   parasites, 65, 66
   viral infections, 17

Z

Zoonotic agents, 85
   bacteria, 42, 45, 47, 51, 53, 54
   parasites, 62, 64
   protozoa, 18
   viruses, 20, 21, 30, 31